W9-AFD-288

Jordan operator algebras

Jordan Operator Algebras

Harald Hanche-Olsen and Erling Størmer

Institute of Mathematics
University of Oslo

Pitman Advanced Publishing Program

Boston · London · Melbourne

PITMAN PUBLISHING LIMITED
128 Long Acre, London WC2E 9AN

PITMAN PUBLISHING INC.
1020 Plain Street, Marshfield, Massachusetts 02050

Associated Companies
Pitman Publishing Pty Ltd, Melbourne
Pitman Publishing New Zealand Ltd, Wellington
Copp Clark Pitman, Toronto

First published 1984

AMS Subject Classifications: Main 46L05
Subsidiary 17CXX

British Library Cataloguing in Publication Data
Hanche-Olsen, Harald
Jordan operator algebras.—(Monographs and studies in mathematics; v. 21)
1. Jordan algebras
I. Title II. Størmer, Erling
III. Series
512′.55 QA326

ISBN 0-273-08619-7

Library of Congress Cataloging in Publication Data
Hanche-Olsen, Harald.
Jordan operator algebras.

Bibliography: p.
Includes index.
1. C*-algebras. 2. Jordan algebras. 3. Operator algebras. I. Størmer, Erling. II. Title.
QA326.H36 1984 512′.55 83-23674

ISBN 0-273-08619-7

Filmset and printed in Northern Ireland at The Universities Press (Belfast) Ltd., and bound at the Pitman Press, Bath, Avon.

Preface

Jordan algebras were first studied by Jordan, von Neumann and Wigner in the mid-1930s with the aim of being a suitable setting for axiomatic quantum mechanics. Later on, however, the subject became mainly a branch of algebra, and it was not until the mid-1960s that Jordan algebras were systematically studied from the point of view of functional analysis. From then on there has been developed a theory which closely resembles that of C^* and von Neumann algebras, and which is concerned with the infinite-dimensional analogues of the original algebras of Jordan, von Neumann and Wigner. It is the purpose of this monograph to present this theory of what is now called JB and JBW algebras. We shall assume only the essentials of functional analysis and then develop the theory to a stage from which we hope it should not be too hard to extend to Jordan algebras most results in C^* and von Neumann algebras which may be needed and have natural formulations in a Jordan algebraic setting.

At the present time it is not clear how central JB algebras will be in future mathematics. They have obvious potential interest in applications to C^* algebras, the mathematical foundations of quantum physics and complex functions in several and an infinite number of variables. To be more specific let us indicate some of these applications.

In C^* algebras order theoretic questions are closely related to Jordan algebras, the classical result being that of Kadison [72] stating that an order automorphism of a C^* algebra is a Jordan automorphism. Direct applications of the theory developed in this monograph have so far been made to the study of the state space, antihomomorphisms and of positive idempotent maps of C^* algebras.

In the mathematical foundations of quantum physics one of the natural axioms is that the observables form a Jordan algebra. If we furthermore want the observables to satisfy the functional calculus of spectral theory, we would assume them to form a JB algebra. This connection is thoroughly discussed in the book of Emch [8], and we refer the reader to that book for further details. We should admit, though, that at the present time C^* algebras are more commonly used than JB algebras as a setting for the observables. The reason for this is mainly pragmatic; C^* algebras are more widely known and, as we shall see, JB algebras are so close to

C^* algebras that the phenomena we want to describe algebraically are often well enough described in terms of C^* algebras.

The relationship of Jordan algebras to holomorphic functions in several variables was first noted by Koecher [77], a result which was later extended to an infinite number of variables by Kaup [73]. The key result in this connection states that certain symmetric domains (in $\mathbb{C}^n$ or a complex Banach space) can be completely characterized in terms of JB algebras. This will be made more explicit in Section 3.9 when we have the necessary terminology at hand. We shall not study this connection because it would carry us too far from our main goal, namely to present the theory of JB algebras in a compact and comprehensible form. The interested reader should consult the notes of Loos [10] for a treatment of the finite-dimensional case, and the papers of Kaup and collaborators [32, 76] for the infinite-dimensional case.

We have divided the book into seven chapters. The first recollects facts from functional analysis which will be needed in the sequel. The reader is supposed to know the basic facts on Banach and Hilbert spaces. We then state the main theorems on topological vector spaces and use the book of Dunford and Schwartz [7] as our reference. Complete proofs will be given of easy consequences of the stated theorems and of results not found in Dunford and Schwartz [7]. The only exception is the spectral theorem characterizing Abelian C^* algebras as continuous functions on locally compact Hausdorff spaces. For this the reader must consult other books.

In the second chapter we develop the algebraic theory of Jordan algebras up to the point where we can prove the classical result of Jordan, von Neumann and Wigner, assuming only the basics of algebra. Parts of the theory are, however, quite technical, so we advise the reader on his first reading not to get too involved with some of the proofs.

In the remaining chapters we study the main topic of the book, namely JB algebras and their weakly closed analogues JBW algebras. In Chapters 3 and 4 we develop the basic techniques for JB and JBW algebras respectively. Chapter 5 contains the key technical lemmas which make the subsequent structure theorems possible. These theorems are proved in Chapters 6 and 7. In Chapter 6 we consider the case of spin factors, which are quite different from the other Jordan algebras, studied in Chapter 7. If we combine the results of the two latter chapters we obtain in particular a complete infinite-dimensional extension of the classical result of Jordan, von Neumann and Wigner.

We are greatly indebted to K. McCrimmon for communicating to us the proof of Macdonald's theorem which we present in Chapter 2.

Oslo
January 1984

H. H.-O.
E.S.

Contents

1

Preliminaries in functional analysis

1.1. Topological vector spaces

1.1.1. Throughout this section E will be a real vector space. E' will denote the real vector space of real linear functionals on E. If E has a topology in which the vector space operations are continuous we call E a *topological vector space* and denote by E^* its *dual space*—the real vector space of continuous functionals in E'. We say E is Hausdorff and locally convex if the topology is Hausdorff and each point has a neighbourhood basis consisting of convex open sets. The main result is the geometric form of the Hahn–Banach theorem.

1.1.2. The Hahn–Banach theorem. *Let E be a Hausdorff locally convex topological vector space. Suppose A and B are two non-empty closed convex subsets of E with B compact. Then there exist $\rho \in E^*$ and $\lambda \in \mathbb{R}$ such that $\rho(x) > \lambda$ for $x \in A$ and $\rho(y) < \lambda$ for $y \in B$.*

For a proof see e.g. Dunford and Schwartz [7, V.2.11].

1.1.3. Let E be a Hausdorff locally convex topological vector space. A subset V of E is called a *cone* if $x \in V$ and $\lambda \in \mathbb{R}$, $\lambda > 0$, implies $\lambda x \in V$, and $(-V) \cap V = \{0\}$, where $-V = \{-x \colon x \in V\}$. If V is a convex subset we talk about a convex cone.

1.1.4. Corollary. *Let E be a Hausdorff locally convex topological vector space. Suppose V is a closed convex cone in E. If $E = V - V$ and $x \in E$, $x \notin V$, then there is $\rho \in E^*$ such that $\rho(x) < 0$ and $\rho(V) = \mathbb{R}^+$—the nonnegative real numbers.*

Proof. By the Hahn–Banach theorem (1.1.2) there are $\rho \in E^*$ and $\lambda \in \mathbb{R}$ such that $\rho(x) < \lambda$ and $\rho(y) > \lambda$ for all $y \in V$, letting $A = V$ and $B = \{x\}$ in the theorem. Since $0 \in V$, $\lambda < 0$, hence $\rho(x) < 0$. Thus $\rho(V)$ is a cone in $\mathbb{R}$

not containing all negative numbers, hence $\rho(V)=\mathbb{R}^+$ or $\rho(V)=\{0\}$. But $E=V-V$ and $\rho\neq 0$, so $\rho(V)=\mathbb{R}^+$.

1.1.5. Another consequence of the Hahn–Banach theorem is the Krein–Milman theorem, stated below. If E is a Hausdorff locally convex topological vector space and K a convex subset of E, we say a point $x\in K$ is an *extreme point* of K if $x=ty+(1-t)z$, y, $z\in K$, $t\in(0,1)$, implies $y=z=x$. If $X\subset E$ we write conv X for the set of convex combinations of elements in X, i.e. the smallest convex subset of E containing X. The closure of conv X is denoted by $\overline{\text{conv}}\, X$.

1.1.6. The Krein–Milman theorem. *Let E be a Hausdorff locally convex topological vector space and K a compact convex subset of E. Let ∂K denote the set of extreme points of K. Then $K=\overline{\text{conv}}\,\partial K$.*

For a proof see e.g. Dunford and Schwartz [7, V.8.4].

1.1.7. Let E be a real vector space. If $x\in E$ and $\rho\in E'$ we sometimes write $\rho(x)$ in the dual form $\rho(x)=\langle x,\rho\rangle$. Then it is apparent that E becomes a subspace of $(E')'$ by $\rho\to\langle x,\rho\rangle$. If F is a subspace of E' we then have $E\subset F'$. We say two vector spaces E and F are in *strict duality* if there is given a bilinear form $E\times F\to\mathbb{R}$ such that for $x\in E$, $\langle x,\rho\rangle=0$ for all $\rho\in F$ implies $x=0$, and symmetrically for $\rho\in F$, $\langle x,\rho\rangle=0$ for all $x\in E$ implies $\rho=0$. Then we can think of E as a subspace of F', identifying $x\in E$ with $\rho\to\langle x,\rho\rangle$. We can also think of F as a subspace of E'. We define the *$\sigma(E,F)$ topology* on E as the weakest in which all the functionals in F are continuous. With this topology E becomes a Hausdorff locally convex topological vector space. Similarly we define the $\sigma(F,E)$ topology on F.

1.1.8. Proposition. *Let E and F be real vector spaces in strict duality. If E is given the $\sigma(E,F)$ topology then $E^*=F$.*

For a proof see e.g. Dunford and Schwartz [7, V.3.9].

1.1.9. Let E and F be vector spaces in strict duality. We define the *polar* X^0 of a subspace $X\subset E$ as the set $X^0=\{\rho\in F\colon \rho(x)=0 \text{ for all } x\in X\}$. Then X^0 is a subspace of F.

1.1.10. The bipolar theorem. *Let E and F be real vector spaces in strict duality. Let X be a $\sigma(E,F)$-closed subspace of E. Then $X=X^{00}$.*

Proof. Clearly $X\subset X^{00}$. Suppose $x\in E$, $x\notin X$. By 1.1.8 the dual of E in the $\sigma(E, F)$ topology is F. By the Hahn–Banach theorem (1.1.2) applied to $A = X$ and $B = \{x\}$ there exist $\rho \in F$ and $\lambda \in \mathbb{R}$ such that $\rho(x) > \lambda$ and $\rho(y) < \lambda$ for all $y \in X$. Since X is a subspace $\rho(y) = 0$ for all $y \in X$, i.e. $\rho \in X^0$, and $\lambda > 0$. Thus $x \notin X^{00}$, proving the theorem.

1.1.11. Let E be a normed real vector space, also called a normed linear space in the sequel. With the norm topology E becomes a Hausdorff locally convex topological vector space with dual space E^*. E^* is a Banach space with norm defined by

$$\|\rho\| = \sup_{x \in E_1} |\rho(x)|,$$

where we always denote by E_1 the unit ball $\{x \in E : \|x\| \leq 1\}$ in E. A variation of the geometric Hahn–Banach theorem is the analytic Hahn–Banach theorem.

1.1.12. The Hahn–Banach theorem, analytic form. *Let E be a normed linear space and X a subspace. If $\rho \in X^*$ then there exists $\bar{\rho} \in E^*$ such that its restriction $\bar{\rho} \mid X = \rho$ and $\|\bar{\rho}\| = \|\rho\|$.*

For a proof see e.g. Dunford and Schwartz [7, II.3.9].

1.1.13. Corollary. *Let E be a normed linear space and $x \in E$. Then there exists $\rho \in E^*$ with $\|\rho\| = 1$ such that $\rho(x) = \|x\|$.*

Proof. Let $X = \mathbb{R}x$ and define $\rho' \in X^*$ by $\rho'(\lambda x) = \lambda \|x\|$. Then $\|\rho'\| = 1$, and the desired ρ is found by 1.1.12.

1.1.14. Let E be a normed linear space and X a subspace. Then the quotient space E/X becomes a real vector space. If X is closed in E then E/X is a normed linear space under the *quotient norm*

$$\|x + X\| = \inf_{y \in X} \|x + y\|.$$

1.1.15. Lemma. *Let E be a normed linear space and X a subspace. If $\rho \in E^*$ then $\|\rho \mid X\| = \|\rho + X^0\|$.*

Proof. By the Hahn–Banach theorem (1.1.12) there is $\omega \in E^*$ such that $\|\omega\| = \|\rho \mid X\|$ and $\omega \mid X = \rho \mid X$. Then $\omega - \rho \in X^0$, and so

$$\|\rho \mid X\| = \|\omega\| = \|\rho + (\omega - \rho)\| \geq \|\rho + X^0\|.$$

On the other hand, if $\eta \in X^0$ then $\|\rho + \eta\| \geq \|(\rho + \eta) \mid X\| = \|\rho \mid X\|$, whence

$$\|\rho + X^0\| = \inf_{\eta \in X^0} \|\rho + \eta\| \geq \|\rho \mid X\|.$$

1.1.16. If E is a normed linear space it follows from 1.1.13 that E and E^* are in strict duality. We call the $\sigma(E^*, E)$ topology on E^* the *w^* topology* or the *weak-$*$ topology.* We shall often apply this topology to the second dual E^{**} of E defined as the dual of the Banach space E^*. In our applications the reference space E will be clearly given in the context, so in order to make the distinction between E^* and E^{**} clear we shall use the name w^* topology for the $\sigma(E^*, E)$ topology on E^* and the name weak-$*$ topology for the $\sigma(E^{**}, E^*)$ topology on E^{**}.

An important consequence of the Tychonoff theorem is the Alaoglu theorem.

1.1.17. The Alaoglu theorem. *Let E be a normed linear space. Then the unit ball E_1^* of E^* is w^* compact.*

For a proof see e.g. Dunford and Schwartz [7, V.4.2].

1.1.18. Each element x in a normed linear space E defines an element $\hat{x} \in E^{**}$ by the formula $\hat{x}(\rho) = \rho(x)$, $\rho \in E^*$. In the norm on E^{**} as the dual of E^* we have

$$\|\hat{x}\| = \sup\{|\hat{x}(\rho)|: \rho \in E_1^*\} = \sup\{|\rho(x)|: \rho \in E_1^*\} \leqslant \|x\|.$$

However, from 1.1.13 $\|x\| = \sup\{|\rho(x)|: \rho \in E_1^*\}$, so that $\|\hat{x}\| = \|x\|$. Thus the map $x \to \hat{x}$ is an isometric imbedding of E into E^{**}. We shall often consider E as a subspace of E^{**} in this embedding.

1.1.19. Proposition. *Let E be a normed linear space. Then we have*
(i) *E is weak-$*$ dense in E^{**},*
(ii) *E_1 is weak-$*$ dense in E_1^{**}.*

Proof. Clearly (ii) implies (i). Suppose $x \in E_1^{**}$. If x does not belong to the weak-$*$ closure of E_1 there are by the Hahn–Banach theorem (1.1.2) a weak-$*$ continuous linear functional ρ on E^{**} and $\lambda \in \mathbb{R}$ such that $\rho(x) > \lambda$ and $\rho(y) < \lambda$ for $y \in E_1$. By 1.1.8 $\rho \in E^*$. Since $0 \in E_1$ and $-y \in E_1$ whenever $y \in E_1$, we have $\lambda > 0$ and $-\lambda < \rho(y) < \lambda$ for all $y \in E$. Let $\omega = \lambda^{-1}\rho$. Then $\omega \in E_1^*$. But $\omega(x) > 1$, which is impossible since $x \in E_1^{**}$. This proves the proposition.

1.1.20. Let E and F be normed linear spaces and $\Phi: E \to F$ a bounded linear map. Then its *adjoint map* $\Phi^*: F^* \to E^*$ is defined by $\langle x, \Phi^*(\rho)\rangle = \langle \Phi(x), \rho\rangle$ when $x \in E, \rho \in F^*$. It is clear that Φ^* is linear. It is bounded since

$$\|\Phi^*(\rho)\| = \sup_{x\in E_1} |\langle x, \Phi^*(\rho)\rangle| = \sup_{x\in E_1} |\langle \Phi(x), \rho\rangle|$$
$$\leq \|\rho\| \sup_{x\in E_1} \|\Phi(x)\| = \|\rho\|\,\|\Phi\|.$$

In particular $\|\Phi^*\| \leq \|\Phi\|$. Conversely, if $\varepsilon > 0$ there is $x \in E_1$ such that $\|\Phi\| \leq \|\Phi(x)\| + \varepsilon$. By 1.1.13 there is $\rho \in F_1^*$ such that $\langle \Phi(x), \rho\rangle = \|\Phi(x)\|$. Thus we have

$$\|\Phi\| \leq \|\Phi(x)\| + \varepsilon = \langle \Phi(x), \rho\rangle + \varepsilon \leq \|\Phi^*(\rho)\| + \varepsilon \leq \|\Phi^*\| + \varepsilon,$$

and so $\|\Phi\| = \|\Phi^*\|$.

It is clear from its definition that Φ^* is continuous when both F^* and E^* are given the w^* topology.

1.1.21. Lemma. *Let E and F be normed linear spaces and $\Phi: E \to F$ a bounded linear map. Then there is a unique extension $\Phi^{**}: E^{**} \to F^{**}$ of Φ which is weak-$*$ to weak-$*$ continuous and satisfies $\|\Phi^{**}\| = \|\Phi\|$.*

Proof. We define Φ^{**} as the adjoint map of $\Phi^*: F^* \to E^*$. By 1.1.20 $\|\Phi^{**}\| = \|\Phi^*\| = \|\Phi\|$. Since the w^* topology on E^{**} is the weak-$*$ topology, Φ^{**} is by 1.1.20 weak-$*$ to weak-$*$ continuous. Since by 1.1.19 E is weak-$*$ dense in E^{**}, Φ^{**} is uniquely defined as soon as it is shown that Φ^{**} extends Φ. But this is immediate, since if $x \in E$ then $\langle \Phi^{**}(x), \rho\rangle = \langle x, \Phi^*(\rho)\rangle = \langle \Phi(x), \rho\rangle$.

1.1.22. Corollary. *Let E be a normed linear space and $\rho \in E^*$. Then there is a unique weak-$*$ continuous extension $\bar{\rho}$ of ρ in E^{***}, and $\|\bar{\rho}\| = \|\rho\|$.*

Proof. $\rho: E \to \mathbb{R}$ and $\mathbb{R}^{**} = \mathbb{R}$, so the result is immediate from 1.1.21.

1.1.23. Proposition. *Let E and F be normed linear spaces and $\Phi: E \to F$ a linear isometry. Then $\Phi^{**}: E^{**} \to F^{**}$ is an isometry such that $\Phi^{**}(E^{**})$ is weak-$*$ closed in F^{**}.*

Proof. By the Hahn–Banach theorem (1.1.12) applied to $\Phi(E) \subset F$ it is immediate that $\Phi^*: F^* \to E^*$ is surjective. Let N be the null space of Φ^*. By 1.1.15 and the fact that Φ is an isometry we have for $\rho \in F^*$ that

$$\|\Phi^*(\rho)\| = \sup_{x\in E_1} |\langle \Phi(x), \rho\rangle| = \|\rho \mid \Phi(E)\| = \|\rho + \Phi(E)^0\|,$$

where $\Phi(E)^0$ is the polar of $\Phi(E)$ in F^*. But if $\eta \in F^*$ then $\eta \in \Phi(E)^0$ if and only if $0=\langle\Phi(x),\eta\rangle=\langle x,\Phi^*(\eta)\rangle$ for all $x\in E$, or $\Phi(E)^0=N$. Thus $\|\Phi^*(\rho)\|=\|\rho+N\|$.

Let $y\in E^{**}$ and $\varepsilon>0$. Then there is $\omega\in E_1^*$ such that $\|y\|\leq|y(\omega)|+\varepsilon$. Since Φ^* is surjective there is by the previous paragraph $\rho\in F^*$ such that $\omega=\Phi^*(\rho)$ and $\|\rho\|<1+\varepsilon$. Since by 1.1.21 $\|\Phi^{**}\|=1$ we therefore have

$$\begin{aligned}\|y\|\geq\|\Phi^{**}(y)\|&>(1+\varepsilon)^{-1}\,|\langle\Phi^{**}(y),\rho\rangle|\\&=(1+\varepsilon)^{-1}\,|\langle y,\Phi^*(\rho)\rangle|=(1+\varepsilon)^{-1}\,|\langle y,\omega\rangle|\\&>(1+\varepsilon)^{-1}(\|y\|-\varepsilon).\end{aligned}$$

Since ε is arbitrary $\|y\|=\|\Phi^{**}(y)\|$, and Φ^{**} is an isometry.

Finally let x belong to the weak-$*$ closure of $\Phi^{**}(E^{**})$. Let (y_α) be a net in E^{**} such that $\Phi^{**}(y_\alpha)\to x$ weak-$*$. Then for all $\rho\in F^*$,

$$\langle y_\alpha,\Phi^*(\rho)\rangle=\langle\Phi^{**}(y_\alpha),\rho\rangle\to\langle x,\rho\rangle.$$

By the first part of the proof Φ^* is surjective, hence the net (y_α) converges weak-$*$ to an element $y\in E^{*\prime}$ defined by

$$\langle y,\Phi^*(\rho)\rangle=\lim_\alpha\langle y_\alpha,\Phi^*(\rho)\rangle=\langle x,\rho\rangle.$$

By the first paragraph of the proof, $\|\Phi^*(\rho)\|=\|\rho+N\|$. Thus, given $\sigma\in E^*$ and $\varepsilon>0$ there is $\rho\in F^*$ with $\sigma=\Phi^*(\rho)$ and $\|\rho\|<\|\sigma\|+\varepsilon$. Then $|\langle y,\sigma\rangle|=|\langle x,\rho\rangle|\leq\|x\|\,(\|\sigma\|+\varepsilon)$. Thus y is bounded, i.e. $y\in E^{**}$, $\Phi^{**}(y)=x$, and $x\in\Phi^{**}(E^{**})$, proving that $\Phi^{**}(E^{**})$ is weak-$*$ closed.

1.1.24. We shall in the sequel be much concerned with Banach spaces which have preduals, where we say E_* is a *predual* for the Banach space E if E_* is a Banach space such that $(E_*)^*=E$. For example E^{**} has E^* as its predual. The next result will be useful when we want to show that certain Banach spaces have preduals.

1.1.25. Proposition. *Let E be a Banach space and suppose $E_*\subset E^*$ is a norm-closed subspace such that $E=(E_*)^*$ in the natural duality. Let $M\subset E$ be a $\sigma(E,E_*)$-closed subspace, and let $M_*=\{\phi\mid M\colon\phi\in E_*\}$. Then we have:*

(i) *M_* is a norm-closed subspace of M^*.*
(ii) *For all $\psi\in M_*$ and $\varepsilon>0$ there is $\phi\in E_*$ such that $\psi=\phi\mid M$ and $\|\phi\|\leq\|\psi\|+\varepsilon$.*
(iii) *$M=(M_*)^*$ in the natural duality.*

Proof. E and E_* are in strict duality, so we may as in 1.1.9 let $M^0=\{\phi\in E_*\colon\phi\mid M=0\}$ be the polar of M in E_*. Since M is $\sigma(E,E_*)$-closed the bipolar theorem (1.1.10) shows $M=M^{00}=\{x\in E\colon\phi(x)=0$ for all $\phi\in M^0\}$. The restriction map $E_*\to M_*$ by $\phi\to\phi\mid M$ is linear of norm

≤ 1 and its null space is M^0. Therefore it induces a linear map $E_*/M^0 \to M_*$ by $\phi + M^0 \to \phi \mid M$. By 1.1.15 this map is isometric. Thus M_* is complete, since E_*/M^0 is. This proves (i) and also (ii), since (ii) just reflects the definition of the quotient norm.

To show (iii) note that by the natural duality between E and E_* the evaluation $\phi \to \phi(x)$ at x defines an element $\phi_x \in (M_*)^*$ for each $x \in M$. We show this map is an isometry between M and $(M_*)^*$. Since $E = (E_*)^*$ it follows from (ii) that

$$\begin{aligned}\|\phi_x\| &= \sup\{|\psi(x)|: \psi \in M_*, \|\psi\| < 1\} \\ &= \sup\{|\phi(x)|: \phi \in E_*, \|\phi\| < 1\} \\ &= \|x\|,\end{aligned}$$

so the map $x \to \phi_x$ is isometric. It remains to show it is surjective. For this let $\rho \in (M_*)^*$, and define $\Phi \in (E_*)^*$ by $\Phi(\phi) = \rho(\phi \mid M)$, $\phi \in E_*$. Then $\Phi \mid M^0 = 0$. Since $E = (E_*)^*$ there is $x \in E$ such that $\Phi(\phi) = \phi(x)$, $\phi \in E_*$. Since $\Phi \mid M^0 = 0$, $x \in M^{00} = M$. Finally, if $\phi \in E_*$ then

$$\rho(\phi \mid M) = \Phi(\phi) = \phi(x) = (\phi \mid M)(x) = \phi_x(\phi \mid M),$$

proving that $\rho = \phi_x$. This proves (iii).

1.2. Order unit spaces

1.2.1. A *partially ordered vector space* is a real vector space A with a proper convex cone A^+, sometimes written A_+. We write $a \geq b$, or $b \leq a$, if $a - b \in A^+$. An element $e \in A^+$ is called an *order unit* for A if for all $a \in A$ there is $\lambda > 0$ such that $-\lambda e \leq a \leq \lambda e$. We say A is *Archimedean* if $na \leq e$ for all $n \in \mathbb{N}$—the natural numbers—implies $a \leq 0$. In this case A has a norm given by

$$\|a\| = \inf\{\lambda > 0: -\lambda e \leq a \leq \lambda e\}.$$

We shall often call this norm the *order norm*. A is said to be an *order unit space* if A has an order unit and A is Archimedean. If A is furthermore a Banach space with respect to the order norm then A is called a *complete order unit space*. Note that since each $a \in A$ can be written in the form $a = \|a\|\, e - (\|a\|\, e - a)$, $A = A^+ - A^+$.

If A is an order unit space then an element $\rho \in A^*$ is said to be *positive*, written $\rho \geq 0$, if $\rho(a) \geq 0$ for all $a \in A^+$. The set of positive functionals in A^* is clearly a w^*-closed convex cone, denoted by A^*_+. The set of $\rho \in A^*_+$ such that $\rho(e) = 1$ form a w^*-closed convex set denoted by $S(A)$ and called the *state space* of A. Elements of $S(A)$ are called *states*. The extreme points of $S(A)$ are called *pure states*.

Note that $\rho \in S(A)$ is pure if and only if whenever $\omega \in A_+^*$ and $\rho - \omega \in A_+^*$, i.e. $\omega \leqslant \rho$, then $\omega = \omega(e)\rho$. Indeed, the necessity is trivial, for if $\rho = \lambda\sigma + (1-\lambda)\omega$, with $\sigma, \omega \in S(A)$, then both $\lambda\sigma \leqslant \rho$ and $(1-\lambda)\omega \leqslant \rho$. Conversely, if ρ is pure and $\omega \leqslant \rho$ with $\lambda = \omega(e) \neq 0, 1$, then $\rho = \lambda(\lambda^{-1}\omega) + (1-\lambda)[(1-\lambda)^{-1}(\rho - \omega)]$ is a convex sum of states, hence $\lambda^{-1}\omega = \rho$.

1.2.2. Lemma. *Let A be an order unit space with order unit e. If $\rho \in A^*$ then $\rho \in S(A)$ if and only if $\|\rho\| = \rho(e) = 1$.*

Proof. If $\rho \in S(A)$ and $\|a\| \leqslant 1$ then $e \pm a \geqslant 0$, hence $0 \leqslant \rho(e \pm a) = 1 \pm \rho(a)$, so that $|\rho(a)| \leqslant 1$, proving that $\|\rho\| = 1$. Conversely, if $\|\rho\| = \rho(e) = 1$ let $a \geqslant 0$ with $\|a\| \leqslant 1$. Then $\|e - a\| \leqslant 1$, so that $1 - \rho(a) = \rho(e - a) \leqslant 1$, and so $\rho(a) \geqslant 0$.

1.2.3. Corollary. *If A is an order unit space then we have:*
(i) *$S(A)$ is a w^* compact subset of A_1^*.*
(ii) *$S(A) = \overline{\text{conv}}\, P(A)$, where $P(A)$ is the set of pure states.*

Proof. By 1.2.2 $S(A)$ is a w^*-closed subset of A_1^*, which is w^* compact by the Alaoglu theorem (1.1.17). This proves (i). (ii) is then immediate from the Krein–Milman theorem (1.1.6).

1.2.4. Lemma. *Let A be an order unit space with order unit e. Let B be a subspace of A containing e, and let $B^+ = B \cap A^+$. Then B is an order unit space, and if $\rho \in S(B)$ then there is $\bar{\rho} \in S(A)$ such that $\bar{\rho} \mid B = \rho$. Furthermore, if ρ is pure then $\bar{\rho}$ can be chosen as a pure state of A.*

Proof. It is clear that B is an order unit space with order unit e. If $\rho \in S(B)$ there is by the Hahn–Banach theorem (1.1.12) $\bar{\rho} \in A^*$ such that $\bar{\rho} \mid B = \rho$ and $\|\bar{\rho}\| = \|\rho\| = 1$. Since $\bar{\rho}(e) = \rho(e) = 1$, $\bar{\rho} \in S(A)$ by 1.2.2.

Suppose ρ is a pure state of B. Let $K = \{\phi \in S(A) \colon \phi \mid B = \rho\}$. Then K is a non-empty w^*-closed convex subset of $S(A)$. In particular K is compact so has by the Krein–Milman theorem (1.1.6) an extreme point $\bar{\rho}$. Then $\bar{\rho}$ is a pure state of A. Indeed, if $\phi, \omega \in S(A)$ and $\bar{\rho} = \lambda\phi + (1-\lambda)\omega$, $\lambda \in (0, 1)$, then $\rho = \lambda\phi \mid B + (1-\lambda)\omega \mid B$, so $\rho = \phi \mid B = \omega \mid B$ since ρ is pure on B. Thus $\phi, \omega \in K$. Since $\bar{\rho}$ is an extreme point of K, $\bar{\rho} = \phi = \omega$, proving that $\bar{\rho}$ is a pure state.

1.2.5. Lemma. *Let A be an order unit space and $a \in A$. Then we have:*
(i) *$a \geqslant 0$ if and only if $\rho(a) \geqslant 0$ for all $\rho \in S(A)$.*
(ii) $\|a\| = \sup_{\rho \in S(A)} |\rho(a)|$.
(iii) *There exists a pure state ρ such that $|\rho(a)| = \|a\|$.*

Proof. We first note that A^+ is norm-closed in A. For this let $a \in A$, $a \notin A^+$. Then there is $t > 0$ in $\mathbb{R}$ such that $a + te \notin A^+$, since otherwise $-a \leq se$ for all $s > 0$, and so $-a \leq 0$ by Archimedicity. If $b \in A$ and $\|a - b\| < t$ then $a - b > -te$, so that $b < a + te$, and $b \notin A^+$, proving the assertion.

Let $a \in A$, $a \notin A^+$. By 1.1.4 there is a positive ρ in A^* such that $\rho(a) < 0$. Then $\rho(e)^{-1}\rho \in S(A)$ and takes a negative value on a. This proves the nontrivial half of (i).

To show (ii) suppose $|\rho(a)| \leq 1$ for all $\rho \in S(A)$. Then $\rho(e \pm a) \geq 0$ for all $\rho \in S(A)$, hence $e \pm a \geq 0$ by (i). Thus $\|a\| \leq 1$. The converse is immediate from 1.2.2.

To show (iii) note that by compactness of $S(A)$ there is by (ii) a state ω such that $|\omega(a)| = \|a\|$. Considering $-a$ if necessary there is no restriction to assume $\omega(a) = \|a\|$. Let $K = \{\rho \in S(A): \rho(a) = \|a\|\}$. By the same argument as in the proof of 1.2.4 we find that an extreme point of K is a pure state satisfying (iii).

1.2.6. Lemma. *Let A be an order unit space. Let $-S(A) = \{\rho \in A^*: -\rho \in S(A)\}$. Then we have:*

(i) $A_1^* = \text{conv}[S(A) \cup -S(A)]$.

(ii) *If $\rho \in A^*$ then there are $\sigma, \omega \in A_+^*$ such that $\rho = \sigma - \omega$, and $\|\rho\| = \|\sigma\| + \|\omega\|$.*

Proof. Since by 1.2.3 $S(A)$ is a w^*-compact convex subset of A_1^* the convex hull $\text{conv}[S(A) \cup -S(A)]$ is a w^*-compact convex subset of A_1^*. If it is a proper subset there exist by the Hahn–Banach theorem (1.1.2) a w^*-continuous linear functional a on A^* and $\rho \in A_1^*$ such that

$$a(\rho) > \sup\{a(\omega): \omega \in \text{conv}[S(A) \cup -S(A)].$$

By 1.1.8 and the identification of A as a subspace of A^{**}, $a \in A$. Thus by 1.2.5 we have

$$\rho(a) = a(\rho) > \sup_{\omega \in S(A)} |\omega(a)| = \|a\|,$$

contrary to the choice of $\rho \in A_1^*$. Thus (i) follows.

By the above, if $\rho \in A^*$ and $\|\rho\| = 1$ then there exist $\sigma', \omega' \in S(A)$ and $\lambda \in [0, 1]$ such that $\rho = \lambda\sigma' - (1 - \lambda)\omega'$. If $\sigma = \lambda\sigma'$ and $\omega = (1 - \lambda)\omega'$, then $\sigma, \omega \in A$, $\|\sigma\| = \lambda$, $\|\omega\| = 1 - \lambda$, $\rho = \sigma - \omega$ and $\|\rho\| = 1 = \|\sigma\| + \|\omega\|$. This proves (ii) when $\|\rho\| = 1$. For a general nonzero ρ in A^* an application of the preceding to $\|\rho\|^{-1}\rho$ completes the proof.

1.2.7. Lemma. *Let A be an order unit space with order unit e. Then A^{**} is a complete order unit space with order unit e and positive cone given by*

$$A_+^{**} = \{a \in A^{**}: a(\rho) \geq 0 \text{ for all } \rho \in S(A)\}.$$

Furthermore, the order norm on A^{**} coincides with the norm obtained as the dual space of A^*.

Proof. Clearly A_+^{**} is a convex cone. If $\rho \in S(A)$ and $a \in A^{**}$ then by 1.2.2 $|a(\rho)| \leq \|a\| \, \|\rho\| = \|a\| \, \rho(e) = \|a\| \, e(\rho)$, so that e is an order unit for A^{**}, and $\| \;\; \|$ majorizes the order norm $\| \;\; \|_0$. If $na \leq e$ for $n \in \mathbb{N}$ then $na(\rho) \leq 1$ for all $\rho \in S(A)$, so that $a \leq 0$, hence A^{**} is Archimedean. If $a \in A^{**}$ then

$$\|a\|_0 = \inf\{\lambda > 0 \colon \lambda e(\rho) \geq |a(\rho)| \text{ for all } \rho \in S(A)\}.$$

Since by 1.2.6

$$\|a\| = \sup_{\rho \in S(A)} |a(\rho)|,$$

$\|a\| \leq \|a\|_0$, so the two norms coincide. Since A^{**} is the dual of A^* it is complete with respect to $\| \;\; \|$, hence it is a complete order unit space.

1.2.8. Corollary. *Let A be an order unit space. Then we have:*
(i) *A^+ is weak-$*$ dense in A_+^{**}.*
(ii) *$A^+ \cap A_1$ is weak-$*$ dense in $A_+^{**} \cap A_1^{**}$.*

Proof. The map $a \to \frac{1}{2}(a+e)$ is an affine isomorphism of A mapping the unit ball $A_1 = \{a \in A \colon -e \leq a \leq e\}$ onto the positive unit ball $A^+ \cap A_1 = \{a \in A \colon 0 \leq a \leq e\}$. By 1.2.7 A^{**} is an order unit space, so the same statement holds for A^{**}. Thus (ii) follows from the weak-$*$ density of A_1 in A_1^{**} (1.1.19).

(i) is an obvious consequence of (ii), for if $a \in A_+^{**}$ then $\|a\|^{-1} a \in A_+^{**} \cap A_1^{**}$.

1.2.9. If A and B are order unit spaces with order units e and f respectively we say an injective linear map $\alpha \colon A \to B$ is an *order isomorphism* if $\alpha(e) = f$ and $a \in A^+$ if and only if $\alpha(a) \in B^+$. An example of such a map is the inclusion map $A \to A^{**}$. This is a consequence of 1.2.5 and 1.2.7.

1.2.10. If K is a convex set we denote by $A^b(K)$ the vector space of bounded real affine functions on K. With pointwise ordering and order unit the constant function 1, $A^b(K)$ becomes a complete order unit space.

1.2.11. Proposition. *Let A be an order unit space. Then the restriction map $a \to a \,|\, S(A)$ is an isometric order isomorphism of A^{**} onto $A^b(S(A))$.*

Proof. Let α denote the restriction map $\alpha(a)=a \mid S(A)$. Then clearly α maps A^{**} into $A^b(S(A))$. By definition of A^{**} (1.2.7), α is clearly an order isomorphism, and since by 1.2.7

$$\|a\| = \sup_{\rho \in S(A)} |a(\rho)|$$

for $a \in A^{**}$, α is an isometry. In order to show α is surjective let $b \in A^b(S(A))$. If $\rho, \omega \in S(A)$ and $\lambda \in [0,1]$ let $a[\lambda\rho-(1-\lambda)\omega]= \lambda b(\rho)-(1-\lambda)b(\omega)$. Since b as a function on $S(A)$ is affine, a is by 1.2.6 a well defined bounded affine functional on A_1^*, and $a(0)=0$. Put $a(\lambda\rho)= \lambda a(\rho)$ for $\lambda>0$. Then $a \in A^{**}$ and $a \mid S(A)=b$.

1.3. C^* algebras

1.3.1. Much of the theory developed later will be direct generalizations of the theory of C^* and von Neumann algebras. Technically we shall not need much of that theory as we shall develop the machinery needed as we go along. In this section and the next we shall state the main definitions and, except for the spectral theorem, prove the theorems that will be needed later.

1.3.2. Let $\mathcal{A}$ be an associative algebra over a field Φ which is either $\mathbb{R}$ or $\mathbb{C}$, and assume $\mathcal{A}$ is a Banach space with respect to a norm $\| \ \|$. $\mathcal{A}$ is called a *Banach algebra* if $\|xy\| \leq \|x\|\,\|y\|$ for all $x, y \in \mathcal{A}$. $\mathcal{A}$ is called a *Banach $*$ algebra* if furthermore $\mathcal{A}$ has an isometric involution $x \to x^*$, i.e. $(\lambda x)^* = \bar{\lambda}x^*$, $(xy)^* = y^*x^*$, and $\|x^*\| = \|x\|$ for $\lambda \in \Phi$, $x, y \in \mathcal{A}$. If in addition $\Phi = \mathbb{C}$ and the norm satisfies $\|x^*x\| = \|x\|^2$ for all $x \in \mathcal{A}$ then $\mathcal{A}$ is said to be a *C^* algebra*. If $\mathcal{A}$ is a Banach $*$ algebra then an element $x \in \mathcal{A}$ is called *self-adjoint* if $x = x^*$, and we denote by $\mathcal{A}_{sa}$ the real subspace of $\mathcal{A}$ consisting of self-adjoint elements. x is called *positive*, written $x \geq 0$, if x is of the form $x = y^*y$ with $y \in \mathcal{A}$. If $\mathcal{A}$ is unital with identity 1 the *spectrum*, $\mathrm{Sp}\, x$, of x is the set of $\lambda \in \Phi$ such that $x - \lambda 1$ is not invertible in $\mathcal{A}$.

There are two main examples of C^* algebras. If H is a complex Hilbert space then $B(H)$, the bounded linear operators on H, is a C^* algebra in the usual operator norm

$$\|x\| = \sup_{\|\xi\| \leq 1} \|x\xi\|,$$

and involution x^* defined by $(x\xi, \eta) = (\xi, x^*\eta)$ for all $\xi, \eta \in H$. Note that the C^* axioms hold in $B(H)$, for if $x \in B(H)$ then

$$\|x^*\|\,\|x\| \geq \|x^*x\| \geq \sup_{\|\xi\| \leq 1} (x^*x\xi, \xi) = \|x\|^2,$$

so that $\|x^*\| \geq \|x\|$, whence by symmetry $\|x^*\| = \|x\|$, and so $\|x^*x\| = \|x\|^2$. In particular, if x and y are self-adjoint we have the inequality

$$\|x^2+y^2\| \geq \sup_{\|\xi\|\leq 1} ((x^2+y^2)\xi, \xi) \geq \sup_{\|\xi\|\leq 1} (x^2\xi, \xi) = \|x^2\|.$$

It follows from the above that each norm-closed subalgebra of $B(H)$, closed in the $*$ operation, is a C^* algebra.

The other examples are represented as continuous functions. If X is a locally compact Hausdorff space we shall denote by $C_0^{\mathbb{C}}(x)$ (resp. $C_0(X)$) the set of continuous complex (resp. real) functions on X vanishing at ∞. If X is compact we understand this to mean all continuous complex (resp. real) functions on X, and we shall often write $C(X)$ instead of $C_0(X)$ in this case. It is clear that with pointwise multiplication, $*$ operation $f^*(\omega) = \overline{f(\omega)}$, and norm $\|f\| = \sup_{\omega\in X} |f(\omega)|$, $C_0^{\mathbb{C}}(X)$ is an Abelian C^* algebra with self-adjoint part $C_0(X)$. Furthermore, if X is compact and $f \in C_0^{\mathbb{C}}(X)$ then $\operatorname{Sp} f = f(X)$.

Conversely, if $\mathscr{A}$ is an Abelian Banach $*$ algebra let X denote the set of continuous homomorphisms ω of $\mathscr{A}$ on Φ such that $\omega(x^*) = \overline{\omega(x)}$, and let X have the relative w^* topology from $\mathscr{A}^*$. We call X the *spectrum* of $\mathscr{A}$. Then each $x \in \mathscr{A}$ defines a continuous function $\hat{x}$ on X with values in Φ by the formula $\hat{x}(\omega) = \omega(x)$. The main result we shall need is the abstract spectral theorem.

1.3.3. Theorem. *Let $\mathscr{A}$ be an Abelian C^* algebra. Then the spectrum X of $\mathscr{A}$ is a locally compact Hausdorff space, which is compact if and only if $\mathscr{A}$ is unital. The map $x \to \hat{x}$ is an isometric isomorphism of $\mathscr{A}$ onto $C_0^{\mathbb{C}}(X)$.*

For the proof see any book on C^* algebras, e.g. Dixmier [6, 1.4.1]. The transformation $x \to \hat{x}$ is usually called the *Gelfand transform*. In specific cases the spectrum can be described more precisely.

1.3.4. Proposition. *Let $\mathscr{A}$ be a unital C^* algebra and x a self-adjoint element in $\mathscr{A}$. Let $C^*(x)$ denote the C^* subalgebra of $\mathscr{A}$ generated by x and 1. Then there is a canonical isometric isomorphism of $C^*(x)$ onto $C^{\mathbb{C}}(\operatorname{Sp} x)$.*

Proof. Since $C^*(x)$ is an Abelian unital C^* algebra we may identify it with $C_0^{\mathbb{C}}(X)$ with X a compact Hausdorff space as in 1.3.3. If $\omega \in X$ the value $\omega(x)$ belongs to $\operatorname{Sp} x$ because $\omega(x - \omega(x)1) = 0$, so that $x - \omega(x)1$ belongs to a maximal ideal in $C^*(x)$ and therefore is not invertible. Let $\psi : X \to \operatorname{Sp} x$ denote the map $\psi(\omega) = \omega(x)$. We show that ψ is a homomorphism. Since X has the relative w^* topology from $C^*(x)^*$, it is clear that ψ is continuous. Similarly ψ is injective, for if $\omega, \omega' \in X$ and

$\omega(x)=\omega'(x)$, then ω and ω' coincide on polynomials in x, hence they are equal. Note that since X is compact a function $f\in C_0^{\mathbb{C}}(X)$ is invertible if and only if $f(\omega)\neq 0$ for all $\omega\in X$. Thus if $\lambda\in \mathrm{Sp}\, x$ then there exists $\omega\in X$ such that $\omega(x-\lambda 1)=0$, or $\omega(x)=\lambda$, whence ψ is surjective. Since X is compact ψ is thus a homomorphism, as asserted. But then the map $f\to f\circ\psi^{-1}$ is an isometric isomorphism of $C_0^{\mathbb{C}}(X)$ onto $C_0^{\mathbb{C}}(\mathrm{Sp}\, x)$, proving the proposition.

1.3.5. Note that if ϕ is the isomorphism of $C^*(x)_{\mathrm{sa}}$ onto $C_0(\mathrm{Sp}\, x)$ obtained in 1.3.4 then each $f\in C_0(\mathrm{Sp}\, x)$ defines an element $f(x)\in C^*(x)$ by $\phi(f(x))=f$. In particular if ι denotes the identity function $\lambda\to\lambda$ in $C_0(\mathrm{Sp}\, x)$ then $\phi(x)=\iota$. This construction will be very useful in order to construct operators with special properties.

We next pursue the study of $C_0(X)$ a little further.

1.3.6. Lemma. *Let X be a compact Hausdorff space and I a norm-closed ideal in $C(X)$. Let $Y=\{t\in X: f(t)=0$ for all $f\in I\}$. Then we have:*
(i) *$I=\{f\in C(X): f\mid Y=0\}$,*
(ii) *$C(X)/I$ with the quotient norm is isometrically isomorphic to $C(Y)$.*

Proof. Let $J=\{f\in C(X): f\mid Y=0\}$. Then clearly J is a norm-closed ideal in $C(X)$ containing I. From its definition it is clear that Y is a closed subset of X. Thus its complement Y^c in X is open and in particular a locally compact Hausdorff space. Since each $f\in J$ annihilates Y it is easily seen that the restriction map $f\to f\mid Y^c$ is an isometry of J onto $C_0(Y^c)$. Let $s\neq t$ in Y^c. Then by definition of Y there exists $f\in I$ such that $f(s)\neq 0$. If $f(s)=f(t)$ let $g\in C(X)$ satisfy $g(t)\neq 1$ and $g(s)=1$. Then $h=fg\in I$ and satisfies $h(s)\neq h(t)\neq 0$. By the locally compact version of the Stone–Weierstrass theorem, which follows from the compact version [7, IV.6.16] by taking the one-point compactification, $I=C_0(Y^c)$, or $I=J$, proving (i).

To show (ii) let $\phi: C(X)\to C(Y)$ be the restriction map, $\phi(f)=f\mid Y$. By (i) $\ker\phi=I$, so there exists an injective homomorphism $\alpha: C(X)/I\to C(Y)$ such that $\alpha(f+I)=\phi(f)$. If $f\in C(Y)$ the Tietze's extension theorem [7, I.5.3] implies the existence of $g\in C(X)$ such that $\phi(g)=f$ and $\|g\|=\|f\|$. Thus ϕ, and hence α, is surjective.

To show α is isometric let $h\in C(X)$. Then $\|h\|\geq\|\phi(h)\|$, so if $j\in I$ then $\|h+j\|\geq\|\phi(h+j)\|=\|\phi(h)\|$, whence $\|h+I\|\geq\|\phi(h)\|$. However, by the preceding paragraph there is $g\in C(X)$ such that $\phi(g)=\phi(h)$ and $\|g\|=\|\phi(h)\|$. Then $h-g\in I$, and $\|h+I\|\leq\|h+(g-h)\|=\|g\|=\|\phi(h)\|$, proving that $\|h+I\|=\|\phi(h)\|$, hence that α is an isometry.

1.3.7. Lemma. *Let X and Y be compact Hausdorff spaces and $\phi: C(X)\to C(Y)$ a homomorphism. Then ϕ satisfies the following:*
(i) $\|\phi(f)\|\leqslant\|f\|$ *for all* $f\in C(X)$.
(ii) *If ϕ is injective then ϕ is an isometry.*
In particular each homomorphism of $C(X)$ into $\mathbb{R}$ is continuous.

Proof. We observe that the spectrum of $C_0(X)$ is X, i.e. every nonzero homomorphism $C_0(X)\to\mathbb{R}$ is of the form $f\to f(x)$ for some $x\in X$. Indeed, given such a homomorphism then its null space is a maximal ideal, which by 1.3.6(i) must be of the form $\{f\in C(X): f(x)=0\}$ for some $x\in X$.

From this it follows that given $y\in Y$, $f\to\phi(f)(y)$ is of the form $f\to f(\alpha(y))$, i.e. we get a map $\alpha: Y\to X$ such that $\phi(f)(y)=f(\alpha(y))$ for all $f\in C(X)$. Then (i) is immediate.

Clearly α is continuous. Thus $\alpha(Y)$ is a compact, hence closed, subset of X. If $\alpha(Y)\neq X$ there is $f\in C(X)$, $f\neq 0$ but $f\mid\alpha(Y)=0$. Then $\phi(f)=0$. Thus, if ϕ is injective then α is onto, and then clearly ϕ is an isometry.

1.3.8. Let H be a complex (resp. real) Hilbert space and $B(H)$ the bounded complex (resp. real) linear operators on H. Note that if H is real the complexification $H^{\mathbb{C}}=H+\mathrm{i}H$ is a complex Hilbert space in the inner product

$$(\xi+\mathrm{i}\eta, \xi'+\mathrm{i}\eta')=(\xi,\xi')+(\eta,\eta')+\mathrm{i}(\eta,\xi')-\mathrm{i}(\xi,\eta').$$

Then the imbedding $H\to H^{\mathbb{C}}$ is an isometry which induces an isometric isomorphism of $B(H)$ into $B(H^{\mathbb{C}})$. Thus we may in most cases restrict attention to complex Hilbert spaces.

1.3.9. Let $\mathcal{A}$ be a Banach $*$ algebra over a field Φ (which as before is $\mathbb{R}$ or $\mathbb{C}$). A $*$ *representation* of $\mathcal{A}$ is a Φ-linear homomorphism $\pi:\mathcal{A}\to B(H)$ for a complex Hilbert space H, such that $\pi(x^*)=\pi(x)^*$ for all $x\in\mathcal{A}$. If π is injective we say it is *faithful*. A *left ideal* in $\mathcal{A}$ is a linear subspace N such that $a\in N$, $b\in\mathcal{A}$ implies $ba\in N$.

1.3.10. Proposition. *Let $\mathcal{A}$ be a unital Banach $*$ algebra over Φ such that $\mathcal{A}_{\mathrm{sa}}$ is a complete order unit space with cone $\mathcal{A}_{\mathrm{sa}}^+=\{x^*x: x\in\mathcal{A}\}$ and order unit 1. Let ρ be a Φ-linear functional on $\mathcal{A}$ such that $\rho(x^*)=\overline{\rho(x)}$ for all $x\in\mathcal{A}$ and $\rho\mid\mathcal{A}_{\mathrm{sa}}$ is a state. Then we have:*
(i) *ρ satisfies the Cauchy–Schwarz inequality*

$$|\rho(x^*y)|\leqslant\rho(x^*x)\rho(y^*y),\qquad x,y\in\mathcal{A}.$$

(ii) *There are a complex Hilbert space H_ρ, a unit vector ξ_ρ in H_ρ and a $*$*

representation π_ρ *of* $\mathcal{A}$ *on* H_ρ *such that* $\rho(x)=(\pi_\rho(x)\xi_\rho, \xi_\rho)$ *for all* $x \in \mathcal{A}$.

(iii) *There are a complex Hilbert space H and a* $*$ *representation* π *of* $\mathcal{A}$ *on H which is isometric on* $\mathcal{A}_{sa}$.

Proof. (i) If $\lambda \in \Phi$, $x, y \in \mathcal{A}$ then

$$0 \leq \rho((\lambda x+y)^*(\lambda x+y)) = |\lambda|^2 \rho(x^*x) + 2 \operatorname{Re} \lambda\rho(y^*x) + \rho(y^*y).$$

From this it is clear that if $\rho(x^*x)=0$ then $\rho(y^*x)=0$, and so (i) holds. If $\rho(x^*x) \neq 0$ let $\lambda = -\overline{\rho(y^*x)}\rho(x^*x)^{-1}$, and again (i) follows.

(ii) The set $N=\{x \in \mathcal{A}: \rho(x^*x)=0\}$ is by (i) a norm-closed left ideal in $\mathcal{A}$. The quotient space $\mathcal{A}/N$ is a pre-Hilbert space over Φ with inner product $(x+N, y+N)=\rho(y^*x)$. Let K_ρ be the Hilbert space completion of $\mathcal{A}/N$ and let ξ_ρ be the unit vector $1+N$. We define a homomorphism π' of $\mathcal{A}$ into the linear operators on $\mathcal{A}/N$ by $\pi(x)(y+N)=xy+N$. Note that if $x, y \in \mathcal{A}$ and $z=x^*x$ then $0 \leq (xy)^*(xy) = y^*x^*xy = y^*zy$, so $z \leq w$ implies $y^*zy \leq y^*wy$. Since $\mathcal{A}_{sa}$ is an order unit space, $x^*x \leq \|x^*x\|\, 1$ for all x, whence $y^*x^*xy \leq \|x^*x\|\, y^*y \leq \|x^*\|\, \|x\|\, y^*y = \|x\|^2 y^*y$. Therefore

$$\|\pi'(x)(y+N)\|^2 = \rho((xy)^*(xy)) \leq \|x\|^2\, \|y+N\|^2,$$

so that $\|\pi'(x)\| \leq \|x\|$. Therefore π' extends to a homomorphism of $\mathcal{A}$ into $B(K_\rho)$. If H_ρ is the complexification of K_ρ in the case when $\Phi=\mathbb{R}$, and $H_\rho = K_\rho$ when $\Phi=\mathbb{C}$, we have by 1.3.8 a $*$ representation π_ρ of $\mathcal{A}$ on H_ρ such that

$$\rho(x)=\rho(1x)=(x+N, 1+N)=(\pi_\rho(x)\xi_\rho, \xi_\rho)$$

for all $x \in \mathcal{A}$. This proves (ii).

(iii) If $\Phi=\mathbb{C}$, $\mathcal{A}=\mathcal{A}_{sa}+i\mathcal{A}_{sa}$, so each state on $\mathcal{A}_{sa}$ has by linearity an extension to a state on $\mathcal{A}$. If $\Phi=\mathbb{R}$ each $x \in \mathcal{A}$ can be written in the form $x=y+z$ with $y \in \mathcal{A}_{sa}$ and $z=\frac{1}{2}(x-x^*)$ skew. Thus a state on $\mathcal{A}_{sa}$ has a canonical extension to $\mathcal{A}$ which annihilates the skew elements. We may therefore by (ii) let

$$H = \bigoplus_{\rho \in S(\mathcal{A}_{sa})} H_\rho$$

be the Hilbert space direct sum of the Hilbert spaces H_ρ, and

$$\pi = \bigoplus_{\rho \in S(\mathcal{A}_{sa})} \pi_\rho$$

be the $*$ representation of $\mathcal{A}$ on H defined by $\pi(x)(\sum \eta_\rho) = \sum \pi_\rho(x)\eta_\rho$. Then $\|\pi(x)\| \leq \|x\|$ for all x since $\|\pi_\rho(x)\| \leq \|x\|$ for all ρ. We assert that π is isometric on $\mathcal{A}_{sa}$. Indeed, let $x \in \mathcal{A}_{sa}$. By 1.2.5 there is a state ρ on $\mathcal{A}_{sa}$ such that $|\rho(x)|=\|x\|$. Thus $\|x\|=|(\pi_\rho(x)\xi_\rho, \xi_\rho)| \leq \|\pi_\rho(x)\| \leq \|\pi(x)\| \leq \|x\|$, so that $\|\pi(x)\|=\|x\|$. This completes the proof.

1.3.11. Remark. It will follow from 3.3.10 below that $\mathcal{A}_{sa}$ is an order unit space with cone $\mathcal{A}_{sa}^+=\{x^2: x\in\mathcal{A}_{sa}\}$. Actually, $x^*x\in\mathcal{A}_{sa}^+$ for any $x\in\mathcal{A}$. This fact, which remained an open problem for several years in the infancy of C^* algebra theory, can be found in any book on C^* algebras. It can also be shown by noting that, as in the proof of 3.7.1 below, we need only prove that $x^*x\leqslant 0$ implies $x=0$. If $x^*x\leqslant 0$ and $\lambda\in\mathbb{R}^+$ then $x^*x-\lambda$ is invertible. Hence so is $xx^*-\lambda$, with $(xx^*-\lambda)^{-1}=\lambda^{-1}[x(x^*x-\lambda)^{-1}x^*-1]$. Thus $xx^*\leqslant 0$, but writing $x=y+iz$, $y, z\in\mathcal{A}_{sa}$ we get $x^*x+xx^*=2(y^2+z^2)\geqslant 0$, which implies $x^*x=xx^*=0$. Then the representation π of 1.3.10 is isometric on $\mathcal{A}$. Indeed, if $x\in\mathcal{A}$ then $\|x\|^2=\|x^*x\|=\|\pi(x^*x)\|=\|\pi(x)^*\pi(x)\|=\|\pi(x)\|^2$. This is the famous Gelfand–Naimark theorem for C^* algebras. We should also remark that the construction of H_ρ, π_ρ and ξ_ρ in 1.3.10(ii) is called the GNS construction after Gelfand, Naimark and Segal.

1.4. Von Neumann algebras

1.4.1. Let H be a complex Hilbert space. If $\xi, \eta\in H$ we define $\omega_{\xi,\eta}\in B(H)^*$ by $\omega_{\xi,\eta}(x)=(x\xi, \eta)$. Let $B(H)_\sim$ denote the vector subspace of $B(H)^*$ generated by the functionals $\omega_{\xi,\eta}$, and let $B(H)_*$ denote the norm closure of $B(H)_\sim$ in $B(H)^*$. For us the most important topology on $B(H)$ will be the *ultraweak* topology, which is by definition the $\sigma(B(H), B(H)_*)$ topology. It should be noted that it is somewhat more common to work with the weak operator topology, defined as the $\sigma(B(H), B(H)_\sim)$ topology. We shall, however, have no contact with this topology.

Note that since $\omega_{\xi,\eta}(x^*)=\overline{\omega_{\eta,\xi}(x)}$ for all $\xi, \eta\in H$, $x\in B(H)$, the $*$ operation is ultraweakly continuous on $B(H)$.

Even though the Banach spaces considered in Section 1.1 were real, we shall still consider some complex Banach spaces, e.g. $B(H)$ and H. The same definitions as in 1.1 still make sense and will be used subsequently.

1.4.2. Let ρ be a linear functional on $B(H)$. Then there exist two linear functionals ρ_1 and ρ_2 on $B(H)$ with real restrictions $\rho_j \mid B(H)_{sa}$ to the self-adjoint part of $B(H)$ such that $\rho=\rho_1+i\rho_2$. Indeed, we may just let $\rho_1(x)=\frac{1}{2}[\rho(x)+\overline{\rho(x^*)}]$ and $\rho_2(x)=(1/2i)[\rho(x)-\overline{\rho(x^*)}]$. Note that since $*$ operation is ultraweakly continuous, if $\rho\in B(H)_*$ so are ρ_1 and ρ_2. If we identify ρ and $\rho\mid B(H)_{sa}$ for a functional ρ such that $\rho\mid B(H)_{sa}$ is real, we have thus shown $B(H)_*=(B(H)_{sa})_*+i(B(H)_{sa})_*$, where $(B(H)_{sa})_*$ is the real Banach space of ultraweakly continuous real functionals on $B(H)_{sa}$. In particular it follows that the ultraweak topology restricted to $B(H)_{sa}$ is the $\sigma(B(H)_{sa}, (B(H)_{sa})_*)$ topology.

1.4.3. Lemma. *Let H be a complex Hilbert space. Then we have:*
(i) *The Banach space $B(H)$ is isometrically isomorphic to $(B(H)_*)^*$ in the natural duality.*
(ii) *$B(H)_{sa}$ is isometrically isomorphic to $((B(H)_{sa})_*)^*$ in the natural duality.*

Proof. Let $x \in B(H)$. Then x defines a bounded linear functional on $B(H)_*$ by $\phi_x(\omega) = \omega(x)$, $\omega \in B(H)_*$. Furthermore if x is self-adjoint then $\phi_x \mid (B(H)_{sa})_*$ is real. Clearly $\|\phi_x\| \leq \|x\|$, and the converse inequality follows from the computation, letting $H_1 = \{\xi \in H: \|\xi\| \leq 1\}$:

$$\begin{aligned}\|x\| &= \sup\{\|x\xi\|: \xi \in H_1\}\\ &= \sup\{|(x\xi, \eta)| = |\omega_{\xi,\eta}(x)|: \xi, \eta \in H_1\}\\ &= \sup\{|\phi_x(\omega_{\xi,\eta})|: \|\omega_{\xi,\eta}\| \leq 1\}\\ &\leq \|\phi_x\|.\end{aligned}$$

Thus $x \to \phi_x$ is an isometry. Note that if x is self-adjoint then $\|x\| = \sup\{|(x\xi, \xi)|: \xi \in H_1\}$, so $\|x\| = \|\phi_x \mid (B(H)_{sa})_*\|$.

Conversely let $\psi \in (B(H)_*)^*$. We define a bounded sesquilinear form on H by

$$\langle \xi, \eta \rangle = \psi(\omega_{\xi,\eta}), \qquad \xi, \eta \in H.$$

By the Riesz representation theorem [7, IV.4.5], $H = H^*$, so it is easy to show the existence of $x \in B(H)$ such that $(x\xi, \eta) = \langle \xi, \eta \rangle$. It follows that $\psi(\omega_{\xi,\eta}) = (x\xi, \eta) = \phi_x(\omega_{\xi,\eta})$ for all $\xi, \eta \in H$. Consequently ψ and ϕ_x coincide on $B(H)_\sim$ and therefore on $B(H)_*$. Note that if ψ is real on $(B(H)_{sa})_*$ then $(x\xi, \xi)$ is real for all $\xi \in H$, so x is self-adjoint. Thus (ii) follows.

1.4.4. Theorem. *Let M be an ultraweakly closed linear subspace of $B(H)_{sa}$. Let $M_* = \{\omega \mid M: \omega \in (B(H)_{sa})_*\}$. Then $M = (M_*)^*$ in the natural duality.*

Proof. By 1.4.3 $((B(H)_{sa})_*)^* = B(H)_{sa}$. Since the ultraweak topology on $B(H)_{sa}$ is the $\sigma(B(H)_{sa}, (B(H)_{sa})_*)$ topology, it follows from 1.1.25 that $M = (M_*)^*$.

1.4.5. A *von Neumann algebra* is a C^* algebra which has a predual, viz. $\mathcal{A} = (\mathcal{A}_*)^*$.

1.4.6. Corollary. *Let $\mathcal{M}$ be an ultraweakly closed C^* algebra of operators on a Hilbert space. Then $\mathcal{M}$ is a von Neumann algebra.*

Proof. Since the $*$ operation is ultraweakly continuous $\mathcal{M}_{sa}$ is ultraweakly closed, hence by the theorem equal to $((\mathcal{M}_{sa})_*)^*$. If $\mathcal{M}_*$ denotes the ultraweakly continuous functionals on $B(H)$ restricted to $\mathcal{M}$ it follows by 1.4.2 that $\mathcal{M}_* = (\mathcal{M}_{sa})_* + i(\mathcal{M}_{sa})_*$. It is thus immediate that $\mathcal{M} = (\mathcal{M}_*)^*$, hence that $\mathcal{M}$ is a von Neumann algebra.

1.4.7. Let $\mathcal{M}$ be a von Neumann algebra acting on a Hilbert space H. Let ξ_0 be a unit vector in H. We say ξ_0 is *cyclic* for $\mathcal{M}$ if the set $\{x\xi_0 : x \in \mathcal{M}\}$ is dense in H. ξ_0 is called *separating* for $\mathcal{M}$ if $x\xi_0 = 0$ implies $x = 0$ for $x \in \mathcal{M}$. Let $\mathcal{M}'$ denote the *commutant* of $\mathcal{M}$, i.e. $\mathcal{M}' = \{x' \in B(H) : x'x = xx' \text{ for all } x \in \mathcal{M}\}$. Then clearly $\mathcal{M}'$ is a C^* algebra (it is in fact a von Neumann algebra by 1.4.6). If ξ_0 is cyclic for $\mathcal{M}$ then ξ_0 is separating for $\mathcal{M}'$. Indeed, if $x' \in \mathcal{M}'$ and $x'\xi_0 = 0$, then $0 = xx'\xi_0 = x'x\xi_0$ for all $x \in \mathcal{M}$. Since ξ_0 is cyclic for $\mathcal{M}$, $x' = 0$ as asserted.

1.4.8. Lemma. *Let $\mathcal{M}$ be a von Neumann algebra acting on a Hilbert space H. Suppose ξ_0 is a unit vector in H cyclic for $\mathcal{M}$ such that the state $\omega(x) = (x\xi_0, \xi_0)$ is pure on $\mathcal{M}$. Then $\mathcal{M}' = \mathbb{C}1$.*

Proof. Since $\mathcal{M}'$ is a C^* algebra it is generated by its positive operators. Let $a' \in \mathcal{M}'$, $0 \leqslant a' \leqslant 1$. Define $\rho \in \mathcal{M}^*$ by $\rho(x) = (a'x\xi_0, \xi_0)$. Then $0 \leqslant \rho \leqslant \omega$ since $0 \leqslant a'x \leqslant x$ whenever $0 \leqslant x \leqslant 1$ and $x \in \mathcal{M}$. By 1.2.1 $\rho = \rho(1)\omega$. In particular, if $x, y \in \mathcal{M}$ we have $(a'x\xi_0, y\xi_0) = \rho(y^*x) = \rho(1)\omega(y^*x) = \rho(1)(x\xi_0, y\xi_0)$. Since ξ_0 is cyclic for $\mathcal{M}$, $a' = \rho(1)1 \in \mathbb{C}1$.

1.5. Comments

The material in this chapter is all very standard, and can be found in several books. In Section 1.1 we have chosen the book by Dunford and Schwartz [7] as our standard reference, and have given proofs of results which have either been easy consequences of basic theorems or have not been found in Dunford and Schwartz [7].

Section 1.2 is perhaps the least familiar in the chapter. Order unit spaces were first studied by Kadison [71], and several of the results are due to him. For more recent treatments see the books of Alfsen [1] and Asimow and Ellis [2].

In our treatment of C^* algebras in Section 1.3 we have put emphasis on the Abelian ones. This is due to the fact that we did not want to refer to anything but the spectral theorem (1.3.3).

Our discussion of von Neumann algebras together with Theorem 1.1.25 is to a great extent taken from the book of Strătilă and Zsidó [12]. Our aim has, as for C^* algebras, been to do as little as possible in order to keep our treatment optimally self-contained.

2

Jordan algebras

2.1. Introduction and preliminaries

2.1.1. In this chapter we shall study the purely algebraic aspects of Jordan algebras. This theory, it turns out, is essentially the same over any field of characteristic not two. Therefore we will fix a field Φ of characteristic not two, and by an *algebra* we shall mean an algebra over Φ. Of course the reader may think of Φ as being $\mathbb{R}$ or $\mathbb{C}$, as this will be the case in the rest of the book.

More precisely by an *algebra* we shall mean a vector space A over Φ, with a bilinear composition law $A \times A \to A$. The product of two elements x and y under this composition will usually be denoted xy, $x \cdot y$ or $x \circ y$. Note that we do *not* assume algebras to be associative or commutative unless explicitly stated, i.e. we do not assume the identities $(xy)z = x(yz)$ or $xy = yx$ to hold.

The field Φ may be equipped with a natural *conjugation*, i.e. a map $\lambda \to \bar{\lambda}$ of Φ into itself such that $\bar{\bar{\lambda}} = \lambda$, $\overline{\lambda + \mu} = \bar{\lambda} + \bar{\mu}$ and $\overline{\lambda\mu} = \bar{\lambda}\bar{\mu}$. If $\Phi = \mathbb{R}$ we take $\bar{\lambda} = \lambda$, while if $\Phi = \mathbb{C}$ we take $\bar{\lambda}$ to be the complex conjugate of λ. An *involution* on an algebra A will then be a map $a \to a^*$ of A into itself satisfying $a^{**} = a$, $(a+b)^* = a^* + b^*$, $(\lambda a)^* = \bar{\lambda} a^*$ and $(ab)^* = b^* a^*$ $(a, b \in A, \lambda \in \Phi)$. An element $a \in A$ such that $a = a^*$ is called *self-adjoint* or *Hermitian*. We also call A a $*$ *algebra*.

2.1.2. For any algebra we can define many of the same notions we have for associative algebras. For example an *ideal* in an algebra A is a subspace J such that $a \in A$, $b \in J$ implies $ab \in J$ and $ba \in J$. Then the quotient A/J has a natural structure of an algebra. A *homomorphism* between algebras is a linear map ϕ such that $\phi(ab) = \phi(a)\phi(b)$. If instead $\phi(ab) = \phi(b)\phi(a)$ we talk of an *antihomomorphism*.

An algebra A is called *unital* if it contains an element 1 such that $1a = a1 = a$ for all $a \in A$. Note that the unit 1 is unique. We shall often call a homomorphism *unital* if it maps the unit to the unit.

2.2. Alternative algebras

2.2.1. An algebra A is called *alternative* if it satisfies the identities

$$x^2y = x(xy), \qquad yx^2 = (yx)x. \tag{2.1}$$

If we introduce the *associator*

$$[x, y, z] = (xy)z - x(yz), \tag{2.2}$$

the identity (2.1) can be rewritten as

$$[x, x, y] = [y, x, x] = 0. \tag{2.3}$$

By substituting z for y and $x + y$ for x we immediately conclude that $[x, y, z]+[y, x, z]=0$ and similarly that $[x, y, z]+[x, z, y]=0$. From this we get $[x, y, x]=-[y, x, x]=0$ or

$$(xy)x = x(yx). \tag{2.4}$$

For this reason we can, and shall, drop the parentheses in an expression like xyx. The above identities mean that the associator $[x, y, z]$ *alternates*, i.e. it is invariant under an even permutation of the variables and gets multiplied by -1 by an odd permutation. This (alternative!) description of the identities (2.1) justifies the term 'alternative algebras'.

2.2.2. The Moufang identities. *The following identities are true in any alternative algebra*

$$(xyx)z = x(y(xz)), \tag{2.5}$$

$$z(xyx) = ((zx)y)x, \tag{2.6}$$

$$(xy)(zx) = x(yz)x. \tag{2.7}$$

Proof. The identity $[x, y, z]+[y, x, z]=0$ can be rewritten as $(xy + yx)z = x(yz)+y(xz)$. Denoting $a \circ b = \frac{1}{2}(ab + ba)$ and $L_a(b) = ab$ in any algebra, this can again be rewritten as $L_{x \circ y} = L_x \circ L_y$. Note now that in any alternative algebra one has

$$xyx = 2x \circ (x \circ y) - x^2 \circ y.$$

Therefore $L_{xyx} = L_x L_y L_x$ and (2.5) is proved.

Equation (2.6) is proved similarly or by using (2.5) in the opposite algebra.

To prove (2.7) we write $(xy)(zx) = -[xy, z, x]+((xy)z)x$ and $x(yz)x = -[x, y, z]x + ((xy)z)x$, to see that (2.7) is equivalent to

$$[z, xy, x] = -[z, x, y]x. \tag{2.8}$$

(Remember that associators alternate.) However, the left-hand side of

(2.8) equals $(z(xy))x-z(xyx)$ and the right-hand side equals $(z(xy))x-((zx)y)x$. It is clear then that (2.6) implies (2.8) and hence (2.7).

2.2.3. Theorem. *An alternative algebra generated by two elements is associative.*

Proof. We shall need a linearized version of (2.8). Substituting $x+w$ for x in (2.8) we obtain

$$[z, xy, w]+[z, wy, x]=-[z, x, y]w-[z, w, y]x. \tag{2.9}$$

Assume now that the alternative algebra A is generated by x and y. We shall consider (non-empty) words $z=u_1\ldots u_n$ (with some arrangement of brackets) where $u_i=x$ or y. The length of the word we call $l(z)=n$. We shall prove by induction that $[a, b, c]=0$ if a, b, c are words and $l(a)+l(b)+l(c)\leqslant n$.

For $n=3$ this is evident from (2.1) and (2.4).

Assume the hypothesis holds for $n\geqslant 3$. We can, and shall, ignore brackets in words of length less than or equal to n. Assume $l(a)+l(b)+l(c)\leqslant n+1$. Two of the words must begin with the same letter, say x. Since associators alternate, we may assume that these two words are b and c. We must consider the following three cases: (i) $b=c=x$, (ii) $b=xb'$, $c=x$, (iii) $b=xb'$, $c=xc'$ where b' (resp. c') denotes a non-empty word. (The case $b=x$, $c=xc'$ can be reduced to case (ii), since associators alternate.) Case (i) is easy, using (2.1). Case (ii) is also easy, since (2.8) implies $[a, xb', x]=-[a, x, b']x$, which vanishes by the induction hypothesis. Finally, in case (iii), using (2.9) and the induction hypothesis,

$$[a, xb', c]+[a, cb', x]=0,$$

but the second term vanishes by case (ii), and so the proof is complete.

2.2.4. Lemma. *There are, up to isomorphism, three two-dimensional real algebras with* 1. *The isomorphism classes are characterized by the existence of a basis* $\{1, x\}$ *satisfying* $x^2=1, 0, -1$ *respectively.*

Proof. Let y be any element not in $\mathbb{R}1$. Then $y^2=\alpha 1+\beta y$ for some α, $\beta\in\mathbb{R}$. We find $(y-\beta/2)^2=(\alpha+\beta^2/4)1$, so we can choose x to be a multiple of $y-\beta/2$. The rest is clear.

2.2.5. In the rest of this paragraph, we shall be concerned with real algebras only. A unital algebra is *quadratic* if, for every element a, a^2 is a linear combination of a and 1. Hence, the subalgebra generated by a and 1 is at most two-dimensional, and is therefore $\mathbb{R}1$ or classified by the above lemma. An algebra is a *division algebra* if it has a unit and every nonzero element has a two-sided inverse.

We shall use the symbol $\mathbb{H}$ for the algebra of quaternions. Its definition will be clear from the proof below. The algebra $\mathbb{O}$ of *octonions*, or Cayley numbers, will be described later. Its definition will be more understandable after the proof of the following result.

2.2.6. Theorem. *Any quadratic, alternative real division algebra is isomorphic to either* $\mathbb{R}$, $\mathbb{C}$, $\mathbb{H}$ *or* $\mathbb{O}$.

Proof. Let D be a quadratic, alternative real division algebra. We may assume $D \neq \mathbb{R}1$. If $a \in D$, $a \notin \mathbb{R}1$, a generates an algebra isomorphic to one of those described in 2.2.4. The first two contain some b such that $b^2 = b$ or $b^2 = 0$, respectively, and $b \neq 0, 1$. This is impossible, however, since then $b = b1 = b(bb^{-1}) = b^2b^{-1} = 1$ or 0, respectively. We conclude that the subalgebra generated by a is isomorphic to $\mathbb{C}$.

An *imaginary unit* is an element $i \in D$ such that $i^2 = -1$. Suppose i, j are imaginary units. Then we have

$$\text{if} \quad ij = ji \quad \text{then} \quad i = \pm j. \tag{2.10}$$

Indeed if $ij = ji$ then by (2.7) $(ij)^2 = (ij)(ji) = ij^2i = -i^2 = 1$. Since ij is contained in a copy of $\mathbb{C}$, $ij = \pm 1$ follows. Multiplying by $-j$ we get $i = \pm j$.

We next show that if i and j are imaginary units then

$$ij + ji \in \mathbb{R}1. \tag{2.11}$$

By the first paragraph of the proof there are an imaginary unit k and α, $\beta \in \mathbb{R}$ such that $ij + ji = \alpha 1 + \beta k$. Notice that i commutes with $ij + ji$, since $i(ij + ji) = -j + iji = (ij + ji)i$. Therefore, if $\beta \neq 0$, $ik = ki$, so (2.10) implies $k = \pm i$. Similarly $k = \pm j$, so $i = \pm j$, and therefore $ij + ji = \pm 2$, contradicting the assumption $\beta \neq 0$. This proves (2.11).

We assert next that the set D_0 of multiples of imaginary units is a subspace of D and that $D = D_0 \oplus \mathbb{R}1$. Indeed, suppose i and j are imaginary units with $i \neq \pm j$, and let α, β be nonzero real numbers. By (2.11) $(\alpha i + \beta j)^2 = -\alpha^2 - \beta^2 + \alpha\beta(ij + ji) \in \mathbb{R}1$, say $(\alpha i + \beta j)^2 = \gamma 1$. If $\gamma \leq 0$ then clearly $\alpha i + \beta j \in D_0$. If $\gamma > 0$ then, since $\alpha i + \beta j$ is contained in a copy of $\mathbb{C}$ and has a positive square, $\alpha i + \beta j \in \mathbb{R}1$, which implies $ij = ji$, or by (2.10) $i = \pm j$, contrary to assumption. Therefore D_0 is a subspace of D. It is clear from the first paragraph of the proof that $D = \mathbb{R}1 \oplus D_0$.

Clearly, then, we can define a positive definite inner product on D_0 by

$$ab + ba = -2(a \mid b)1.$$

The imaginary units are the unit vectors of D_0 and an orthonormal set is a set of anticommuting imaginary units (where by anticommuting we mean $ab=-ba$).

Pick now any imaginary unit i. If $D=\mathbb{R}1\oplus\mathbb{R}i$, the proof is complete. If not, pick an imaginary unit j orthogonal to i. That is, $ij=-ji$. Write $k=ij$. Then it is easy to verify that i, j and k satisfy the following identities:

$$\begin{aligned} ij &= -ji = k, \\ jk &= -kj = i, \\ ki &= -ik = j, \\ i^2 &= j^2 = k^2 = -1. \end{aligned}$$

Clearly then $\{1, i, j, k\}$ is a basis for a subalgebra $\mathbb{H}$ of D. The above identities form the definition of the *quaternions*.

Assume that $D\neq\mathbb{H}$. Then we can pick a new unit vector l in D_0 orthogonal to i, j and k. In other words $l^2=-1$, and l anticommutes with i, j and k. Note that $\mathbb{H}\cap\mathbb{H}l=\{0\}$, for if a, $b\in\mathbb{H}$ and $a=bl$ then since $b^{-1}\in\mathbb{R}1+\mathbb{R}b$, the alternative law (2.1) yields $b^{-1}a=b^{-1}(bl)=(b^{-1}b)l=l$, so $l\in\mathbb{H}$, which is impossible.

Write now $\mathbb{O}=\mathbb{H}\oplus\mathbb{H}l$. We shall show that $\mathbb{O}$ is a subalgebra of D by giving a formula for the product in $\mathbb{O}$.

First, however, we note that we can define a real linear map $a\to\bar{a}$ of D into itself such that $\bar{a}=a$ if $a\in\mathbb{R}1$ and $\bar{a}=-a$ if $a\in D_0$. It is easily checked that $\overline{ab}=\bar{b}\bar{a}$ if a, $b\in\mathbb{H}$. The fact that l anticommutes with i, j and k can now be rewritten as

$$al=l\bar{a}, \qquad a\in\mathbb{H}. \tag{2.12}$$

Applying the third Moufang identity (2.7) we get $(bl)(dl)=(l\bar{b})(dl)=l(\bar{b}d)l=(\overline{\bar{b}d})l^2=-\bar{d}b$, or

$$(bl)(dl)=-\bar{d}b. \tag{2.13}$$

From (2.12), with $a=i$, we find $ili=l$. For $x\in\mathbb{H}$ we can then use the first Moufang identity (2.5) to conclude $lx=(ili)x=i(l(ix))$. Multiplication by i yields $i(lx)=-l(ix)=-(\overline{ix})l=(\bar{x}i)l$ or, using $lx=\bar{x}l$ and substituting $\bar{x}$ for x, $i(xl)=(xi)l$. We get similar formulae with j or k substituted for i, so we find

$$y(xl)=(xy)l \qquad (x, y\in\mathbb{H}). \tag{2.14}$$

Applying this in the opposite algebra D^0 we get $(lx)y=l(yx)$ or, using (2.12), $(\bar{x}l)y=(\overline{yx})l=(\bar{x}\bar{y})l$ or

$$(xl)y=(x\bar{y})l. \tag{2.15}$$

Combining (2.12), (2.13) and (2.14) we have

$$(a+bl)(c+dl)=ac-\bar{d}b+(da+b\bar{c})l \tag{2.16}$$

whenever $a, b, c, d \in \mathbb{H}$. This is the defining identity of the octonion algebra $\mathbb{O}$.

We finally show that D cannot properly contain $\mathbb{O}$. Indeed, if $D \neq \mathbb{O}$, we can repeat the foregoing discussion with $\mathbb{O}$ replacing $\mathbb{H}$, thereby finding an imaginary unit m in D, such that if $a, b, c, d \in \mathbb{O}$ then

$$(a+bm)(c+dm)=ac-\bar{d}b+(b\bar{c}+da)m.$$

To obtain a contradiction, we compute the associator $[i, jm, l]$ in two ways: first,

$$\begin{aligned}[i, jm, l] &= (i(jm))l - i((jm)l)\\ &= ((ji)m)l + i((jl)m)\\ &= -(km)l + ((jl)i)m\\ &= (kl)m - ((ji)l)m\\ &= 2(kl)m,\end{aligned}$$

secondly,

$$\begin{aligned}[i, jm, l] &= -[jm, i, l]\\ &= -((jm)i)l + (jm)(il)\\ &= +((ji)m)l + (j(\overline{il}))m\\ &= -(km)l + (j(li))m\\ &= +(kl)m - ((ij)l)m\\ &= +(kl)m - (kl)m\\ &= 0.\end{aligned}$$

This contradiction completes the proof.

2.3. Special Jordan algebras

2.3.1. Consider any algebra $\mathcal{A}$. If $a, b \in \mathcal{A}$, let

$$a \circ b = \tfrac{1}{2}(ab+ba). \tag{2.17}$$

Then $\circ$ defines a bilinear, commutative product on $\mathcal{A}$. Thus $\mathcal{A}^{\mathrm{J}}$, which by definition is the vector space $\mathcal{A}$ with the product $\circ$, is a commutative algebra. If $\mathcal{A}$ is associative, we call $\circ$ the *special Jordan product* in $\mathcal{A}$. It should be noted that even when $\mathcal{A}$ is associative, $\mathcal{A}^{\mathrm{J}}$ is not, as the following simple example shows. Let $\mathscr{A}=M_2(\Phi)$, and let

$$a=\begin{pmatrix}1&0\\0&0\end{pmatrix},\quad b=\begin{pmatrix}0&0\\0&1\end{pmatrix},\quad c=\begin{pmatrix}0&1\\1&0\end{pmatrix}.$$

Then $(a\circ b)\circ c=0$, while $a\circ(b\circ c)=1/4c$.

It is interesting to note, however, that the product $\circ$, if $\mathscr{A}$ is associative, satisfies the following weak form of associativity:

$$a\circ(b\circ a^2)=(a\circ b)\circ a^2. \tag{2.18}$$

Let $\mathscr{A}$ be an associative algebra. By a *Jordan subalgebra* of $\mathscr{A}$ we mean a subalgebra of $\mathscr{A}^{\mathrm{J}}$, i.e. a linear subspace of $\mathscr{A}$ which is closed under the Jordan product $\circ$ defined in (2.17). Any algebra isomorphic to a Jordan subalgebra of an associative algebra will be called a *special Jordan algebra*.

2.3.2. The *Jordan triple product* is an interesting and useful algebraic combination of elements in an associative algebra which, perhaps surprisingly, can be expressed by the Jordan product. It is defined by

$$\{abc\}=\tfrac{1}{2}(abc+cba). \tag{2.19}$$

Indeed, it is easy to check that the formula

$$\{abc\}=(a\circ b)\circ c+(b\circ c)\circ a-(a\circ c)\circ b \tag{2.20}$$

holds. This formula and the following special case of it appear so often that they are worth remembering:

$$\{aba\}=2a\circ(a\circ b)-a^2\circ b. \tag{2.21}$$

It is natural to ask whether this can be done for more than three variables, i.e. generalize (2.19) to

$$\{a_1\dots a_n\}=\tfrac{1}{2}(a_1\dots a_n+a_n\dots a_1). \tag{2.22}$$

Can this be expressed in terms of the Jordan product, if $n\geqslant 4$? No, there even exist Jordan subalgebras of associative algebras which are not closed under the multilinear product (2.22) if $n=4$. We shall call a Jordan subalgebra *reversible* if it is closed under (2.22) for all $n\in\mathbb{N}$, and *irreversible* otherwise.

2.3.3. The study of free algebras is very important. Here we shall need two of them: the *free (unital) associative algebra* $FA\{x_1,x_2,\dots,x_n\}$ and the *free (unital) special Jordan algebra* $FS\{x_1,x_2,\dots,x_n\}$, which is the Jordan subalgebra of $FA\{x_1,\dots,x_n\}$ generated by $x_1,\dots,x_n$ and 1. In this notation, we have sacrificed any reference to the underlying field Φ in favour of the mnemonics '*FA*' and '*FS*'. This should cause no confusion.

Some remarks should be made at this point. In the above definition $x_1, \ldots, x_n$ should be distinct letters and $FA\{x_1, \ldots, x_n\}$ should be thought of as consisting of formal linear combinations of words composed of the letters $x_1, \ldots, x_n$, including the empty word, denoted by 1. Then words are multiplied by juxtaposition, and the product in $FA\{x_1, \ldots, x_n\}$ is defined by extending this product bilinearly. To gain some familiarity with this algebra the reader should convince himself that the polynomial ring $\Phi[x_1, \ldots, x_n]$ is the quotient of $FA\{x_1, \ldots, x_n\}$ by the ideal generated by all elements $x_ix_j - x_jx_i$, $1 \leq i < j \leq n$.

2.3.4. The universal property of $FS\{x_1, \ldots, x_n\}$. *Let A be a special Jordan algebra with a unit 1. If $y_1, \ldots, y_n \in A$ there is a unique homomorphism $\phi: FS\{x_1, \ldots, x_n\} \to A$ such that $\phi(1) = 1$ and $\phi(x_i) = y_i$, $i = 1, \ldots, n$.*

Proof. We may assume that A is a Jordan subalgebra of an associative algebra $\mathcal{A}$. We may even assume that 1 is a unit of $\mathcal{A}$; otherwise replace $\mathcal{A}$ by $1\mathcal{A}1$.

It is obvious that there exists a homomorphism $\psi: FA\{x_1, \ldots, x_n\} \to \mathcal{A}$ mapping 1 to 1 and x_i to y_i. Indeed, ψ must map the word $x_{i_1} \ldots x_{i_m}$ to $y_{i_1} \ldots y_{i_m}$ and the empty word to 1. Since these words form a basis for $FA\{x_1, \ldots, x_n\}$, ψ exists. Clearly, ψ is also a homomorphism $FA\{x_1, \ldots, x_n\}^J \to \mathcal{A}^J$. Let ϕ be the restriction of ψ to $FS\{x_1, \ldots, x_n\}$. Since $x_1, \ldots, x_n$ and 1 generate $FS\{x_1, \ldots, x_n\}$, the rest is now clear.

2.3.5. A linear combination in $FA\{x_1, \ldots, x_n\}$ of 1 and elements $\{x_{i_1} \ldots x_{i_m}\}$ (see (2.22)) will be called *reversible*. Consider the subspace H of all reversible elements. We can give an alternative description of H as follows. First, note that $FA\{x_1, \ldots, x_n\}$ has a natural involution $*$, defined by $(x_{i_1} \ldots x_{i_m})^* = x_{i_m} \ldots x_{i_1}$ and linearity. If we call an element a *self-adjoint* if $a = a^*$, then the space H of reversible elements coincides with the space of self-adjoint elements of $FA\{x_1, \ldots, x_n\}$, since if $a = a^* \in FA\{x_1, \ldots, x_n\}$ then a, being a linear combination of terms $x_{i_1} \ldots x_{i_m}$, and satisfying $a = \frac{1}{2}(a + a^*)$, is also a linear combination of terms $\frac{1}{2}(x_{i_1} \ldots x_{i_m} + (x_{i_1} \ldots x_{i_m})^*) = \{x_{i_1} \ldots x_{i_m}\}$. It follows that H is a reversible Jordan subalgebra and, indeed, is the smallest reversible Jordan subalgebra containing $x_1, \ldots, x_n$.

By a *tetrad* we shall mean an element $\{x_{i_1}x_{i_2}x_{i_3}x_{i_4}\}$, where $1 \leq i_1 < i_2 < i_3 < i_4 \leq n$.

2.3.6. Theorem. *The space of reversible elements of $FA\{x_1, \ldots, x_n\}$ coincides with the Jordan algebra generated by 1, $x_1, \ldots, x_n$ and the tetrads.*

Before proving this, let us point out a few of its corollaries to see what is going on. First, if $n<4$ there are no tetrads, so we get:

2.3.7. Corollary. *If $n\leqslant 3$ the free special Jordan algebra $FS\{x_1, \ldots, x_n\}$ coincides with the space of reversible elements in $FA\{x_1, \ldots, x_n\}$.*

2.3.8. Corollary. *Assume $\mathscr{A}$ is an associative algebra and A is a Jordan subalgebra generated by at most three elements (and 1, if A has a unit). Then A is reversible.*

Proof. We may assume that A has a unit. (If not, append one.) Then we may assume that 1 is also a unit of $\mathscr{A}$. Let ψ: $FA\{x, y, z\}\to\mathscr{A}$ be a homomorphism mapping 1 to 1 and x, y, z onto a set of generators of A. Then ψ maps $FS\{x, y, z\}$ onto A. By 2.3.5 and 2.3.7, $FS\{x, y, z\}$ is reversible in $FA\{x, y, z\}$. Hence A is reversible in $\mathscr{A}$.

2.3.9. Corollary. *Assume $\mathscr{A}$ is an associative algebra, A a Jordan subalgebra such that $\{a_1a_2a_3a_4\}\in A$ whenever $a_i\in A$. Then A is reversible.*

Proof. Let $a_1, \ldots, a_n\in A$. As before, we may assume A has a unit which is also the unit of $\mathscr{A}$. Let ψ: $FA\{x_1, \ldots, x_n\}\to\mathscr{A}$ be the homomorphism such that $\psi(1)=1$ and $\psi(x_i)=a_i$. By assumption, ψ maps all tetrads into A. Hence ψ maps the reversible element $\{x_1, \ldots, x_n\}$ into A, i.e. $\{a_1\ldots a_n\}=\psi(\{x_1\ldots x_n\})\in A$.

Let us now proceed with the proof.

2.3.10. Proof of 2.3.6. From the remarks preceding the statement of the theorem it is clear that the space H of reversible elements must contain the Jordan algebra A generated by $1, x_1, \ldots, x_n$, and the tetrads.

To complete the proof, we must show that $H\subseteq A$, and for this it suffices to show $\{x_{i_1}\ldots x_{i_m}\}\in A$ for all m. Clearly, it is true for $m\leqslant 3$. Assume it holds for all $m<r$, where $r\geqslant 4$. We shall prove that it holds for $m=r$.

Let us write $\equiv$ for congruence modulo A, i.e. $a\equiv b$ means $a-b\in A$. First, note that

$$2x_{i_1}\circ\{x_{i_2}\ldots x_{i_r}\}=\{x_{i_1}\ldots x_{i_r}\}+\{x_{i_2}\ldots x_{i_r}x_{i_1}\}.$$

By induction the left-hand side of this equation belongs to A. Hence

$$\{x_{i_2}\ldots x_{i_r}x_{i_1}\}\equiv-\{x_{i_1}\ldots x_{i_r}\}. \tag{2.23}$$

Each time the indices are permuted cyclically, the sign of $\{x_{i_1}\ldots x_{i_r}\}$ will change. Hence, if r is odd, in the end we have $\{x_{i_1}\ldots x_{i_r}\}\equiv-\{x_{i_1}\ldots x_{i_r}\}$, or $\{x_{i_1}\ldots x_{i_r}\}\in A$, and the proof is complete.

If r is even, we note that

$$4(x_{i_1}\circ x_{i_2})\circ\{x_{i_3}\ldots x_{i_r}\}=\{x_{i_1}x_{i_2}x_{i_3}\ldots x_{i_r}\}+\{x_{i_2}x_{i_1}x_{i_3}\ldots x_{i_r}\}$$
$$+\{x_{i_3}\ldots x_{i_r}x_{i_1}x_{i_2}\}+\{x_{i_3}\ldots x_{i_r}x_{i_2}x_{i_1}\}.$$

Again, by induction, the left-hand side belongs to A. Hence we get

$$\{x_{i_1}x_{i_2}x_{i_3}\ldots x_{i_r}\}+\{x_{i_2}x_{i_1}x_{i_3}\ldots x_{i_r}\}+\{x_{i_3}\ldots x_{i_r}x_{i_1}x_{i_2}\}$$
$$+\{x_{i_3}\ldots x_{i_r}x_{i_2}x_{i_1}\}\equiv 0. \quad (2.24)$$

Note that the third term in (2.24) comes from the first by applying the cyclic permutation in (2.23) twice; hence the two terms are congruent mod A. Similarly, the second and fourth terms are congruent. Therefore, (2.24) implies

$$\{x_{i_1}\ldots x_{i_r}\}\equiv-\{x_{i_2}x_{i_1}x_{i_3}\ldots x_{i_r}\}. \quad (2.25)$$

Notice that both (2.23) and (2.25) can be expressed as

$$\{x_{i_1}\ldots x_{i_r}\}\equiv(-1)^{\sigma}\{x_{i_{\sigma(1)}}\ldots x_{i_{\sigma(r)}}\}, \quad (2.26)$$

where σ is a permutation of $\{1,\ldots,r\}$ and $(-1)^{\sigma}$ its sign. (Remember that when r is even, an r-cycle is odd.) Since the r-cycle in (2.23) and the transposition in (2.25) generate the whole symmetric group, (2.26) holds for all permutations σ. Since, by assumption, $\{x_{i_1}\ldots x_{i_4}\}\in A$ whenever $i_1<i_2<i_3<i_4$, it follows from (2.26) that $\{x_{i_1}\ldots x_{i_4}\}\in A$ whenever i_1, i_2, i_3, i_4 are *distinct* indices. On the other hand, if any two of the indices $i_1,\ldots,i_r$ are equal, it follows from (2.26) (by applying a transposition) that $\{x_{i_1}\ldots x_{i_r}\}\in A$. This completes the proof when $r=4$.

We may now assume that $r\geqslant 5$. We proceed in a fashion similar to what we did to prove (2.23) and (2.24), and note that

$$4\{x_{i_1}\ldots x_{i_4}\}\circ\{x_{i_5}\ldots x_{i_r}\}=\{x_{i_1}\ldots x_{i_r}\}+\{x_{i_4}\ldots x_{i_1}x_{i_5}\ldots x_{i_r}\}$$
$$+\{x_{i_1}\ldots x_{i_4}x_{i_r}\ldots x_{i_5}\}+\{x_{i_4}\ldots x_{i_1}x_{i_r}\ldots x_{i_5}\}.$$

Once more, the left-hand side belongs to A by induction. The permutation reversing $1,\ldots,4$ is even, so from this and (2.26) we get

$$\{x_{i_1}\ldots x_{i_r}\}\equiv-\{x_{i_1}\ldots x_{i_4}x_{i_r}\ldots x_{i_5}\}.$$

But the sign of the permutation that reverses $5,\ldots,r$ is $(-1)^{r/2}$ (remember r is even), so we obtain from (2.26) and the above that $\{x_{i_1}\ldots x_{i_r}\}\in A$, if r is a multiple of 4.

Finally, if r is not a multiple of 4, the reversal of $(1,\ldots,r)$ is an odd permutation, so (2.26) predicts $\{x_{i_1}\ldots x_{i_r}\}\equiv-\{x_{i_r}\ldots x_{i_1}\}$. However, $\{x_{i_r}\ldots x_{i_1}\}=\{x_{i_1}\ldots x_{i_r}\}$, so once more we have $\{x_{i_1}\ldots x_{i_r}\}\equiv 0$. The proof is complete.

2.3.11. We shall need another consequence of 2.3.6 that lies somewhat deeper than the preceding corollaries. It is known that the homomorphic image of a special Jordan algebra need not be special [9, p. 11].

2.3.12. Lemma. *Let I_z be the ideal generated by z in $FS\{x, y, z\}$, and let J_z be the ideal generated by z in $FA\{x, y, z\}$. Then $I_z = J_z \cap FS\{x, y, z\}$.*

Proof. Clearly $J_z \cap FS\{x, y, z\}$ is an ideal in $FS\{x, y, z\}$ containing z, so $I_z \subseteq J_z \cap FS\{x, y, z\}$.

To prove the opposite inclusion consider the definition of $FA\{x, y, z\}$ given in 2.3.3. If w is a word in x, y, z, let $\partial_z w$ be the number of occurrences of z in that word. Let Z_i be the linear span of all words w with $\partial_z w = i$. Then, trivially,

$$FA\{x, y, z\} = \bigoplus_{i=0}^{\infty} Z_i, \qquad J_z = \bigoplus_{i=1}^{\infty} Z_i. \tag{2.27}$$

(Incidentally, $FA\{x, y\}$ is canonically isomorphic to Z_0.)

A Jordan monomial in x, y, z is any element of $FS\{x, y, z\}$ constructed according to the following rules: 1, x, y and z are Jordan monomials, and so is $a \circ b$ whenever a and b are Jordan monomials. We note the following simple facts. Every Jordan monomial c belongs to some Z_i. If $i \geqslant 1$ then $c \in I_z$. Moreover $FS\{x, y, z\}$ is the linear span of all Jordan monomials.

From this we see that if $u \in FS\{x, y, z\}$ we can write $u = v + w$ where $v \in Z_0$ and $w \in I_z$. If moreover $u \in J_z$ then by (2.27) $v = 0$, so $u = w \in I_z$. In other words, $J_z \cap FS\{x, y, z\} \subseteq I_z$, and the proof is complete.

2.3.13. Theorem. *Any homomorphic image of $FS\{x, y\}$ is a special Jordan algebra.*

Proof. Let I be an ideal in $FS\{x, y\}$. Let J be the ideal in $FA\{x, y\}$ generated by I. We must prove $J \cap FS\{x, y\} = I$. Then, clearly, $FS\{x, y\}/I$ will be isomorphic to a Jordan subalgebra of $FA\{x, y\}/J$, and hence is special.

First, note that J is the linear span

$$J = \text{lin}\{bac\colon a \in I,\, b, c \in FA\{x, y\}\}. \tag{2.28}$$

Denote by $*$ the natural involution in $FA\{x, y\}$, as in 2.3.5. Then, clearly, if $u \in FS\{x, y\}$, $u^* = u$, or $u = \frac{1}{2}(u + u^*)$. By 2.3.7, the converse is also true, and therefore it follows from (2.28) that

$$J \cap FS\{x, y\} = \text{lin}\{bac + c^*ab^*\colon a \in I,\, b, c \in FA\{x, y\}\}.$$

What must be proved, therefore, is that $bac+c^*ab^* \in I$ if $a \in I$ and b, $c \in FA\{x, y\}$.

We shall identify $FA\{x, y\}$ with its canonical image in $FA\{x, y, z\}$. Let ϕ be the homomorphism $FA\{x, y, z\} \to FA\{x, y\}$ mapping 1, x, y, z to $1, x, y$ and a respectively. Then $\phi(bzc+c^*zb^*)=bac+c^*ab^*$. By 2.3.7, the reversible element $bzc+c^*zb^*$ belongs to $FS\{x, y, z\}$. In the notation of 2.3.12, it also belongs to J_z, and therefore it is in I_z. However, ϕ restricts to a homomorphism of $FS\{x, y, z\}$ onto $FS\{x, y\}$, so I_z is clearly mapped into I (because $a \in I$). Hence $bac+c^*ab^* \in \phi(I_z) \subseteq I$, and the proof is complete.

2.4. Jordan algebras

2.4.1. Let A be an algebra with the product written $(a, b) \to a \circ b$. A is called a *Jordan algebra* if the following two identities are satisfied for all $a, b \in A$:

$$a \circ b = b \circ a, \tag{2.29}$$

$$a \circ (b \circ a^2) = (a \circ b) \circ a^2. \tag{2.30}$$

It should be remarked that from (2.18) it immediately follows that any special Jordan algebra is a Jordan algebra. The converse is not true; Jordan algebras which are not special will be called *exceptional*. However, there is more power in the above two axioms than meets the eye. Indeed, consider a Jordan polynomial $p(x, y, z)$ in the three variables x, y, z. Assume that p is of degree at most 1 in z; for example,

$$p(x, y, z) = x^2 \circ (y \circ (y \circ z)) - (x^2 \circ y^2) \circ z.$$

Then *Macdonald's theorem* states that, if $p(x, y, z)=0$ for all x, y, z in a special Jordan algebra, then $p(x, y, z)=0$ for all x, y, z in any Jordan algebra. This is a very effective tool for proving identities. Most of this section is devoted to a proof of Macdonald's theorem.

2.4.2. Let us for the moment turn to a discussion of the axioms (2.29) and (2.30). While axiom (2.29) simply states that A is commutative, axiom (2.30), called the *Jordan axiom*, can profitably be rewritten. We start by defining the *multiplication operator* $T_a: A \to A$ for $a \in A$ by

$$T_a b = a \circ b. \tag{2.31}$$

We prefer the notation T_a to the more customary L_a or R_a to avoid confusion when we work inside an associative algebra. Indeed, in an associative algebra $T_a = \frac{1}{2}(L_a + R_a)$, where L_a (resp. R_a) is left (resp. right) multiplication. Axiom (2.30), with the help of commutativity, now takes the form

$$T_a T_{a^2} = T_{a^2} T_a, \tag{2.32}$$

i.e. T_a and T_{a^2} commute.

Note that from (2.32) it is trivial that we can adjoin an identity to a Jordan algebra and still have a Jordan algebra. Indeed, this follows from the identity

$$[T_{(a+\lambda 1)^2}, T_{(a+\lambda 1)}] = [T_{a^2} + 2\lambda T_a + \lambda^2 \iota, T_a + \lambda \iota] = 0,$$

where $\lambda \in \Phi$ and ι is the identity map.

2.4.3. We shall derive a 'linearized' version of (2.32). For this, substitute $a + \lambda b + \mu c$ for a in $[T_a, T_{a^2}] = 0$, where λ, $\mu \in \Phi$ and a, b, $c \in A$. Expanding, we have a polynomial of degree 3 in λ, μ with coefficients in A. This polynomial vanishes for all values of λ, μ. We would like to conclude that all coefficients of the polynomial vanish. Then we could compute the $\lambda\mu$ term of the polynomial, to arrive at the *linearized Jordan axiom*:

$$[T_a, T_{b\circ c}] + [T_b, T_{c\circ a}] + [T_c, T_{a\circ b}] = 0. \tag{2.33}$$

This is the only point at which the characteristic of Φ gives us a little trouble. Indeed there are polynomials with nonzero coefficients which vanish for nonzero characteristics: for example, $\lambda^3 + 2\lambda = 0$ for all $\lambda \in \mathbb{Z}/3\mathbb{Z}$. However, the above derivation of (2.33) is still valid, for the third-degree terms in our polynomial are $\lambda^3[T_b, T_{b^2}] + \mu^3[T_c, T_{c^2}] = 0$, so the polynomial really has degree at most 2. Since Φ does not have characteristic 2, it does follow that the coefficients of the polynomial vanish. (Just substitute $\lambda = 0, 1, 2$ and similarly for μ.)

In the converse direction we may note that, except possibly in characteristic 3, the linearized Jordan axiom implies the ordinary Jordan axiom.

2.4.4. Applying the linearized Jordan axiom (2.33) to an element $d \in A$ and collecting terms with the same signs on separate sides of the equality sign, we obtain the equivalent form

$$\begin{aligned} &a\circ((b\circ c)\circ d) + b\circ((c\circ a)\circ d) + c\circ((a\circ b)\circ d) \\ &\quad = (b\circ c)\circ(a\circ d) + (c\circ a)\circ(b\circ d) + (a\circ b)\circ(c\circ d). \end{aligned} \tag{2.34}$$

Note that the right-hand side is symmetric in all four variables. In particular, the left-hand side is invariant when a and d are interchanged. This fact, when written in terms of the operator T in order to eliminate the variable d, becomes the formula

$$T_a T_{b\circ c} + T_b T_{c\circ a} + T_c T_{a\circ b} = T_{a\circ(b\circ c)} + T_b T_a T_c + T_c T_a T_b. \tag{2.35}$$

We are now in the position to show that Jordan algebras are *power associative*, that is a subalgebra generated by a single element is associative. To state this differently, define powers inductively by $a^0 = 1$, $a^{n+1} = a \circ a^n$ $(n \geq 1)$. Then we have:

2.4.5. Lemma. *Let A be a Jordan algebra and $a \in A$. Then, for natural numbers m and n,*
(i) $a^{m+n} = a^m \circ a^n$,
(ii) $T_{a^m} T_{a^n} = T_{a^n} T_{a^m}$.

Proof. Let $n \geq 3$ and apply (2.35) with $b = a^{n-2}$, $c = a$. We get a formula expressing $T_{a^n} = T_{a \circ (b \circ c)}$ in terms of the operators T_{a^k}, $1 \leq k \leq n-1$. By induction it follows that T_{a^n} is a polynomial in T_a and T_{a^2}, which commute, so (ii) follows.

To show (i) we use induction on m. By definition (i) holds for $m = 1$. Assuming it holds for some $m \geq 1$, we get by (ii)

$$\begin{aligned} a^n \circ a^{m+1} &= a^n \circ (a^m \circ a) = T_{a^n} T_a a^m = T_a T_{a^n} a^m \\ &= a \circ (a^n \circ a^m) = a \circ a^{n+m} = a^{n+m+1}. \end{aligned}$$

2.4.6. In order to state Macdonald's theorem in its precise form, we must consider free Jordan algebras. First consider the *free nonassociative algebra* $FN\{x_1, \ldots, x_n\}$. It is defined analogously to the free associative algebra, but the insertion of parentheses is now important. To be precise, a nonassociative monomial is a word built from the letters $x_1, \ldots, x_n$ and the special symbols '(' and ')', which can be constructed according to the following recursive rules. The empty word (denoted by 1) and $x_1, \ldots, x_n$ are nonassociative monomials. If a and b are non-empty nonassociative monomials, then so is (ab). It should be noted that if a, b, c, d are non-empty nonassociative monomials and $(ab) = (cd)$, then $a = c$ and $b = d$. We can now define a nonassociative polynomial to be a formal linear combination of nonassociative monomials, and $FN\{x_1, \ldots, x_n\}$ to be the linear space of nonassociative polynomials in $x_1, \ldots, x_n$. $FN\{x_1, \ldots, x_n\}$ is an algebra, with the product defined by

$$\left(\sum_i \lambda_i a_i\right)\left(\sum_j \mu_j b_j\right) = \sum_{ij} \lambda_i \mu_j a_i b_j,$$

where $\lambda_i \mu_j \in \Phi$ and a_i, b_j are monomials (and we write $a_i b_j$ instead of $(a_i b_j)$). The universal property of $FN\{x_1, \ldots, x_n\}$ is summed up in the following:

2.4.7. Lemma. *Let A be a (possibly nonassociative) unital algebra and let $a_1, \ldots, a_n \in A$. Then there is a unique algebra homomorphism of $FN\{x_1, \ldots, x_n\}$ into A mapping 1 to 1 and x_i to a_i, $i = 1, \ldots, n$.*

Proof. The uniqueness is clear, since 1 and $x_1, \ldots, x_n$ generate $FN\{x_1, \ldots, x_n\}$. To define the homomorphism it is enough to define it on the basis of monomials. By definition 1 must be mapped to 1 and x_i to a_i. If w is a monomial other than 1 or x_i, then w can be *uniquely* written $w=(uv)$, with u, v monomials of shorter length than w. If u is mapped to b and v to c, then by induction we can map w to bc. Extending linearly we clearly get a homomorphism.

2.4.8. We can now define the *free Jordan algebra* $FJ\{x_1, \ldots, x_n\}$ generated by n distinct letters $x_1, \ldots, x_n$ (and 1) to be the quotient $FN\{x_1, \ldots, x_n\}/I$ by the ideal I generated by all expressions of the form $ab-ba$ and $(ab)a^2-a(ba^2)$, where a, $b \in FN\{x_1, \ldots, x_n\}$. Clearly $FJ\{x_1, \ldots, x_n\}$ is a Jordan algebra. It satisfies a universal property, given by the following:

2.4.9. Lemma. *Let A be a unital Jordan algebra and $a_1, \ldots, a_n \in A$. There exists a unique homomorphism $FJ\{x_1, \ldots, x_n\} \to A$ mapping 1 to 1 and x_i to a_i $(i=1, \ldots, n)$.*

Proof. As in 2.4.7, uniqueness is clear. The existence follows from 2.4.7, since the canonical homomorphism $FN\{x_1, \ldots, x_n\} \to A$ constructed there must annihilate the generators of I (by the Jordan axioms for A), and therefore factors through $FJ\{x_1, \ldots, x_n\}$.

2.4.10. Of particular interest to us will be the canonical, surjective homomorphism

$$FJ\{x_1, \ldots, x_n\} \to FS\{x_1, \ldots, x_n\}$$

mapping 1 to 1 and x_i to x_i $(i=1, \ldots, n)$. We shall call the kernel of this map the *exceptional ideal* of $FJ\{x_1, \ldots, x_n\}$. From 2.4.5 we see that the exceptional ideal vanishes if $n=1$. Later we shall see that it also vanishes if $n=2$. Macdonald's theorem below can be paraphrased as a statement saying something about the exceptional ideal if $n=3$.

2.4.11. An element p of $FJ\{x_1, \ldots, x_n\}$ may also be called a *Jordan polynomial* in $x_1, \ldots, x_n$. We shall write $p(a_1, \ldots, a_n)$ for $\phi(p)$ if A is a Jordan algebra with 1, $a_i \in A$, and ϕ is the canonical homomorphism $\phi\colon FJ\{x_1, \ldots, x_n\} \to A$ mapping $1 \to 1$ and $x_i \to a_i$.

2.4.12. By a *multiplication operator* on a Jordan algebra A we shall mean a linear operator on A belonging to the algebra generated by all the operators $T_a\colon b \to a \circ b$, where $a \in A$. By a multiplication operator in two variables x, y we shall mean an element of the algebra generated by all operators T_p on $FJ\{x, y, z\}$, where $p \in FJ\{x, y\} \subseteq FJ\{x, y, z\}$. (Note that the homomorphism $FJ\{x, y\} \to FJ\{x, y, z\}$ mapping $1 \to 1$, $x \to x$, $y \to y$ has a right inverse given by $1 \to 1$, $x \to x$, $y \to y$, $z \to 0$. We identify $FJ\{x, y\}$ with its image in $FJ\{x, y, z\}$.) Let M be a multiplication operator in the two variables x, y. We shall see that, by substituting elements of any Jordan algebra A for x and y, we get a multiplication operator on A. For this we define

$$M(a, b)c = \phi(Mz),$$

where $a, b, c \in A$ and $\phi: FJ\{x, y, z\} \to A$ is the homomorphism mapping $1 \to 1$, $x \to a$, $y \to b$, $z \to c$. To see that $M(a, b)$ is indeed a multiplication operator, note first that if $p \in FJ\{x, y\}$ then the above definition becomes (for $M = T_p$)

$$T_p(a, b)c = \phi(p \circ z) = \phi(p) \circ c = p(a, b) \circ c = T_{p(a,b)}c,$$

so $T_p(a, b)$ is indeed the multiplication operator $T_{p(a,b)}$.

Next, if p is a polynomial and M is a multiplication operator then

$$\begin{aligned}(T_pM)(a, b)c &= \phi(T_pMz) = \phi(p \circ Mz) = \phi(p) \circ \phi(Mz) \\ &= p(a, b) \circ M(a, b)c = T_p(a, b)M(a, b)c,\end{aligned}$$

and by induction we get that, for any multiplication operator M in two variables, $M(a, b)$ is indeed a multiplication operator. Moreover if M and M' are different multiplication operators then $(MM')(a, b) = M(a, b)M'(a, b)$. Finally, note that $M(x, y) = M$ if $A = FJ\{x, y, z\}$.

2.4.13. Macdonald's theorem. *Let M be a multiplication operator in two variables x, y. If $M(a, b) = 0$ for all a, b in all special Jordan algebras, then $M = 0$.*

The proof of Macdonald's theorem, being quite technical, is left to the end of this section. We turn now to its applications.

2.4.14. The Shirshov–Cohn theorem. *Any Jordan algebra generated by two elements (and 1, if unital) is special.*

Proof. First, we shall show that $FJ\{x, y\}$ is special. Indeed, we shall prove that the canonical surjection $FJ\{x, y\} \to FS\{x, y\}$ is an isomorphism. So assume $p \in FJ\{x, y\}$ belongs to the kernel of this map (i.e. the exceptional ideal). Then T_p is a multiplication operator in two variables on $FJ\{x, y, z\}$ (which contains $FJ\{x, y\}$, see 2.4.12), which vanishes on $FS\{x, y, z\}$, and therefore on all special Jordan algebras by the universal property of $FS\{x, y, z\}$ (2.3.4). By Macdonald's theorem T_p vanishes on $FJ\{x, y, z\}$, so $p = T_p1 = 0$.

Now, let A be a Jordan algebra generated by two elements (and 1; if A is not unital, append a unit). Then by 2.4.9 A is a quotient of $FJ\{x, y\}$ and so, by the above, of $FS\{x, y\}$. By 2.3.13 A is special.

2.4.15. Macdonald's theorem is usually used in the following rephrased form:

Any polynomial identity in three variables, with degree at most 1 *in the third variable, and which holds in all special Jordan algebras, holds in all Jordan algebras.*

To see that this follows from 2.4.13, note first that 2.4.14 takes care of that part of the identity which does not contain the third variable. Next, note that a Jordan polynomial p in three variables x, y, z, and linear in z, can be written $p = Mz$, where M is a multiplication operator in x, y. Then, if $p(a, b, c) = 0$ for all a, b, c in a special Jordan algebra, it follows that $M(a, b) = 0$ for all a, b in a special Jordan algebra, so by 2.4.13 $M = 0$, and then $p = 0$ follows

2.4.16. We now give two examples illustrating the use of the Shirshov–Cohn and Macdonald theorems. For this we shall need the *Jordan triple product.* It will also be needed in the proof of Macdonald's theorem and, indeed, it will follow us throughout the book. One reason is that it 'behaves better' than the Jordan product, and triple product identities are also easier to visualize than most Jordan product identities. Recall the definition from 2.3 of the triple product,

$$\{abc\} = \tfrac{1}{2}(abc + cba)$$

in any associative algebra. In (2.20) it was noted that this can be expressed in terms of the Jordan product as follows:

$$\{abc\} = (a \circ b) \circ c + (c \circ b) \circ a - (a \circ c) \circ b. \tag{2.36}$$

This, then, will be our *definition* of $\{abc\}$ in a general Jordan algebra. We shall also write $U_{a,c}$ for the operator

$$U_{a,c}(b) = \{abc\}, \tag{2.37}$$

and $U_a = U_{a,a}$. Then we get the following operator identities:

$$U_{a,c} = T_a T_c + T_c T_a - T_{a \circ c}, \tag{2.38}$$

$$U_a = 2T_a - T_{a^2}. \tag{2.39}$$

2.4.17. As an example of the Shirshov–Cohn theorem we have the following identity:

$$\{aba\}^2 = \{a\{ba^2b\}a\}. \tag{2.40}$$

To prove (2.40), note that it clearly holds in any special algebra, but the Shirshov–Cohn theorem states that the algebra generated by a and b is indeed special.

2.4.18. Analogously, we have the identity

$$\{a\{b\{aca\}b\}a\}=\{\{aba\}c\{aba\}\} \tag{2.41}$$

from Macdonald's theorem. Indeed, (2.41) is quite clearly true in any special Jordan algebra and so by 2.4.15 it is true in all Jordan algebras. The multiplication operator M in 2.4.15 will, of course, be $M=U_xU_yU_x-U_{\{xyx\}}$.

2.4.19. We now start our work on the proof of Macdonald's theorem. For the proof we shall need some identities that are valid in all Jordan algebras. These are derived in 2.4.20–2.4.22. It may be interesting to note that they all follow from Macdonald's theorem—this is true even for the four-variable identities in 2.4.20, which can be derived from suitable three-variable identities by linearization.

The main difficulty in proving Macdonald's theorem is that multiplication operators can be written in many ways. We shall resolve this problem by putting the algebra of multiplication operators in two variables into a one-to-one correspondence with a more easily understood space. In 2.4.23 we prove a simple lemma on generators of multiplication operator algebras, which will be needed to show that this correspondence is onto. Lemma 2.4.24 contains all the technical difficulties in establishing this correspondence. We recommend that the proof of 2.4.24, except the first paragraph, is skipped on a first reading. In 2.4.25 we complete the proof of Macdonald's theorem.

2.4.20. Lemma. *For a, b, c, d in any Jordan algebra the following identities hold:*

$$\{abc\}\circ d=\{(a\circ d)bc\}+\{ab(c\circ d)\}-\{a(b\circ d)c\}, \tag{2.42}$$

$$\{abc\}\circ d=\{a(b\circ c)d\}-\{(a\circ c)bd\}+\{c(a\circ b)d\}, \tag{2.43}$$

$$\{abc\}\circ d+\{dbc\}\circ a=\{a(b\circ c)d\}+\{(a\circ d)bc\}. \tag{2.44}$$

Proof. Using the abbreviations

$$(abcd)=(a\circ b)\circ(c\circ d)+(a\circ c)\circ(b\circ d)+(a\circ d)\circ(b\circ c),$$

$$(a;bcd)=(a\circ(b\circ c))\circ d+(a\circ(c\circ d))\circ b+(a\circ(d\circ b))\circ c,$$

we see from (2.34) that $(abcd)=(a;bcd)$. Since $(abcd)$ is invariant under all permutations of a, b, c and d, so is $(a;bcd)$. Now expand the terms in

(2.42):

$$\begin{aligned}
-\{abc\}\circ d &= -(a\circ(b\circ c))\circ d+(b\circ(a\circ c))\circ d-(c\circ(a\circ b))\circ d\\
\{(a\circ d)bc\} &= (a\circ d)\circ(b\circ c) \qquad +(b\circ(a\circ d))\circ c-(c\circ(a\circ d))\circ b\\
-\{a(b\circ d)c\} &= (a\circ c)\circ(b\circ d)-(a\circ(b\circ d))\circ c \qquad -(c\circ(b\circ d))\circ a\\
\{ab(c\circ d)\} &= (a\circ b)\circ(c\circ d)-(a\circ(c\circ d))\circ b+(b\circ(c\circ d))\circ a.
\end{aligned}$$

Adding these formulae, we get on the right-hand side $(abcd)-(a;bcd)+(b;acd)-(c;abd)$, which equals 0 by the above remarks. This proves (2.42). Equations (2.43) and (2.44) are proved similarly.

2.4.21. Lemma. *The following formulae hold in any Jordan algebra, where l, m, n are natural numbers:*

$$2T_{a^l}U_{a^m,a^n}=2U_{a^m,a^n}T_{a^l}=U_{a^{m+l},a^n}+U_{a^m,a^{n+l}}, \tag{2.45}$$

$$U_{a^n}=U_a^n. \tag{2.46}$$

Proof. By 2.4.5 all the operators U_{a^m,a^n} and T_{a^l} commute. In particular the first half of (2.45) follows. Using this and (2.42) we get

$$\begin{aligned}
a^l\circ\{a^mba^n\} &= \{a^{m+l}ba^n\}+\{a^mba^{n+l}\}-\{a^m(b\circ a^l)a^n\}\\
&= \{a^{m+l}ba^n\}+\{a^mba^{n+l}\}-a^l\circ\{a^mba^n\},
\end{aligned}$$

from which (2.45) follows.

Using (2.45) twice we find

$$\begin{aligned}
U_aU_{a^n} &= 2T_a^2U_{a^n}-T_{a^2}U_{a^n}\\
&= 2T_aU_{a^{n+1},a^n}-U_{a^{n+2},a^n}\\
&= U_{a^{n+2},a^n}+U_{a^{n+1},a^{n+1}}-U_{a^{n+2},a^n}\\
&= U_{a^{n+1}},
\end{aligned}$$

from which (2.46) follows by induction.

2.4.22. Lemma. *In any Jordan algebra the following identities hold, where $m\leqslant n$ are natural numbers:*

$$T_{a^n}U_{a^m,b}+T_{a^m}U_{a^n,b}=U_{a^m,a^n}T_b+U_{a^{m+n},b}, \tag{2.47}$$

$$2U_{a^n,b}T_{a^m}=U_{a^{n-m},b}U_{a^m}+U_{a^{m+n},b}, \tag{2.48}$$

$$2T_{a^m}U_{a^n,b}=U_{a^m}U_{a^{n-m},b}+U_{a^{m+n},b}. \tag{2.49}$$

Proof. Equation (2.47) is an immediate consequence of (2.44) with a^m, b, a^n substituted for a, c and d respectively.

Substituting $a=x$, $c=z$, $d=y$ (2.43) can be rearranged and rewritten as

$$U_{x\circ z,y}=U_{x,y}T_z+U_{z,y}T_x-T_yU_{x,z}, \tag{2.50}$$

while $a=x$, $c=y$, $d=z$ in (2.44) similarly yields

$$U_{x\circ z,y}=T_zU_{x,y}+T_xU_{z,y}-U_{x,z}T_y. \tag{2.51}$$

Notice the symmetry between (2.50) and (2.51). We shall use (2.50) to prove (2.48). It will then be left to the reader to see that the same method proves (2.49) from (2.51).

We first substitute x^2 for x in (2.50):

$$U_{x^2\circ z,y}=U_{x^2,y}T_z+U_{z,y}T_{x^2}-T_yU_{x^2,z}.$$

In the first and third terms, use (2.50) with $z=x$ and y (resp. z) substituted for y. In the second term, use $T_{x^2}=2T_x^2-U_x$. We get

$$U_{x^2\circ z,y}=(2U_{x,y}T_x-T_yU_x)T_z+U_{z,y}(2T_x^2-U_x)$$
$$-T_y(2U_{x,z}T_x-T_zU_x).$$

We shall let x and z be powers of a. Therefore, by 2.4.5, $T_xT_z=T_zT_x$ and $U_xT_z=T_zU_x$. If we use this in the above equation and rearrange terms, we get

$$U_{x^2\circ z,y}=2(U_{x,y}T_z+U_{z,y}T_x-T_yU_{x,z})T_x-U_{z,y}U_x,$$

using (2.50) again,

$$U_{x^2\circ z,y}=2U_{x\circ z,y}T_x-U_{z,y}U_x.$$

With $x=a^m$, $z=a^k$, $y=b$ and $n=m+k$, this is formula (2.48). Formula (2.49) is proved similarly, using (2.47).

2.4.23. Lemma. *Let A be a unital Jordan algebra. Let B be a subalgebra of A containing* 1, *and let E be the algebra of multiplication operators on A generated by* $\{T_b: b\in B\}$. *Suppose X is a set of generators for B containing* 1. *Then E is generated by* $\{U_{x,y}: x, y\in X\}$.

Proof. Since $U_{x,1}=T_x$ and $U_{x,y}=T_xT_y+T_yT_x-T_{x\circ y}$, it is enough to show that E is generated by $\{T_{x\circ y}: x, y\in X\}$. In other words if p is a monomial in the elements of X then we must show that T_p is a polynomial in $\{T_{x\circ y}: x, y\in X\}$. This is clear if p is of degree $\leqslant 2$, while if p is of degree $\geqslant 3$ we have $p=a\circ(b\circ c)$, where a, b, c are monomials of lower degree than p. But then (2.35) implies

$$T_p=T_aT_{b\circ c}+T_bT_{c\circ a}+T_cT_{a\circ b}-T_bT_aT_c-T_cT_aT_b,$$

so induction on the degree of p completes the proof.

2.4.24. Lemma. *To each pair p, q of monomials in the free unital associative algebra $FA\{x, y\}$ we can associate a multiplication operator $M_{p,q} = M_{q,p}$ in the two variables x, y such that*

(i) $M_{p,q}z = \frac{1}{2}(pzq^* + qzp^*)$ *in* $FS\{x, y, z\}$,
(ii) $M_{1,1} = 1$,
(iii) $U_{x^k}M_{p,q} = M_{x^kp,x^kq}$, $U_{y^k}M_{p,q} = M_{y^kp,y^kq}$,
(iv) $U_{x^k,y^l}M_{p,q} = \frac{1}{2}(M_{x^kp,y^lq} + M_{y^lp,x^kq})$.

Proof. First note that, by 2.3.7, the map $M_{p,q}$ as defined by (i) really does map $FS\{x, y, z\}$ into itself. Clearly, on $FS\{x, y, z\}$ (ii)–(iv) hold. However, it is not so obvious that $M_{p,q}$ is a multiplication operator. Our task will be to lift $M_{p,q}$ to a multiplication operator on $FJ\{x, y, z\}$, and then to show (ii)–(iv) here.

Our definition will be by induction on the *weight* of a monomial, defined as follows. Any monomial p in x, y can be written $p = r_1r_2 \ldots r_n$, where $n \geq 0$ and each r_i is a positive power of x or y, the powers of x alternating with those of y. The weight of p is then $w(p) = n$. If p starts with an x, i.e. $n \geq 1$ and r_1 is a power of x, we say $p \in X_0$. Similarly, if p starts with a y, we say $p \in Y_0$. We write $X = X_0 \cup \{1\}$, $Y = Y_0 \cup \{1\}$.

Our definition by induction will require many forms of the induction step, according to the form of p, q. We will always present these in symmetric form, so that the equality $M_{p,q} = M_{q,p}$ will be evident. Also, the definition will be symmetric in the letters x, y. We shall introduce the definition of $M_{p,q}$ gradually, proving (i) as we go along.

The basis of induction: define

$$M_{x^i,y^j} = M_{y^j,x^i} = U_{x^i,y^j} \qquad (i \geq 0, j \geq 0). \tag{2.52}$$

It is obvious that (i) holds for the $M_{p,q}$ defined so far. Also, $i = j = 0$ in (2.52) yields (ii).

The following set of definitions is for the case when p, q start with the same letter:

$$M_{x^ip,x^jq} = U_{x^j}M_{x^{i-j}p,q} \qquad (i \geq j \geq 1, p, q \in Y), \tag{2.53a}$$

$$M_{x^ip,x^jq} = U_{x^i}M_{p,x^{j-i}q} \qquad (j \geq i \geq 1, p, q \in Y), \tag{2.53b}$$

$$M_{y^ip,y^jq} = U_{y^j}M_{y^{i-j}p,q} \qquad (i \geq j \geq 1, p, q \in X), \tag{2.54a}$$

$$M_{y^ip,y^jq} = U_{y^i}M_{p,y^{j-i}q} \qquad (j \geq i \geq 1, p, q \in X). \tag{2.54b}$$

Consider (2.53a). If $M_{x^{i-j}p,q}$ satisfies (i), then in $FS\{x, y, z\}$ we have

$$\begin{aligned} M_{x^ip,x^jq}z &= U_{x^j}M_{x^{i-j}p,q}z \\ &= x^j[\tfrac{1}{2}(x^{i-j}pzq^* + qzp^*x^{i-j})]x^j \\ &= \tfrac{1}{2}(x^ipzq^*x^j + x^jqzp^*x^i), \end{aligned}$$

hence (2.53a), and similarly (2.53b), (2.54a) and (2.54b), preserve the truth of (i).

For the case of different first letters, we introduce the following pair of definitions, where $i \geqslant 1$, $j \geqslant 1$, and either $p \neq 1$ or $q \neq 1$:

$$M_{x^ip,y^jq} = 2U_{x^i,y^j}M_{p,q} - M_{y^jp,x^iq} \qquad (p \in Y, q \in X), \tag{2.55a}$$

$$M_{y^jp,x^iq} = 2U_{y^j,x^i}M_{p,q} - M_{x^ip,y^jq} \qquad (p \in X, q \in Y). \tag{2.55b}$$

Now consider (2.55a). If $M_{p,q}$ and M_{y^jp,x^iq} satisfy (i), then so does M_{x^ip,y^jq} by the following calculation, starting from (2.55a):

$$\begin{aligned} M_{x^ip,y^jq}z &= \tfrac{1}{2}x^i(pzq^* + qzp^*)y^j + \tfrac{1}{2}y^j(pzq^* + qzp^*)x^i \\ &\quad - \tfrac{1}{2}(y^jpzq^*x^i + x^iqzp^*y^j) \\ &= \tfrac{1}{2}(x^ipzq^*y^j + y^jqzp^*x^i). \end{aligned}$$

Similarly, (2.55b) preserves the truth of (i).

Notice that in (2.53a)–(2.55b) $M_{r,s}$ is defined in terms of operators $M_{p,q}$ where $w(p) \leqslant w(r)$ and $w(q) < w(s)$, or $w(p) < w(r)$ and $w(q) \leqslant w(s)$. These definitions, and (2.52), cover all the cases where r and s both have weight $\geqslant 1$. To cover the remaining cases we include the following definitions; where $i \geqslant 1$:

$$M_{x^ip,1} = 2T_{x^i}M_{p,1} - M_{p,x^i} \qquad (p \in Y_0), \tag{2.56a}$$

$$M_{1,x^ip} = 2T_{x^i}M_{1,p} - M_{x^i,p} \qquad (p \in Y_0), \tag{2.56b}$$

$$M_{y^ip,1} = 2T_{y^i}M_{p,1} - M_{p,y^i} \qquad (p \in X_0), \tag{2.57a}$$

$$M_{1,y^ip} = 2T_{y^i}M_{1,p} - M_{y^i,p} \qquad (p \in X_0). \tag{2.57b}$$

Actually, (2.56a) is the special case (excluded above) $j=0$, $q=1$ in (2.55a), and therefore by the same calculation preserves the truth of (i).

This finishes the definition of $M_{p,q}$. The reader may not, however, be convinced that we have covered all combinations p, q. To see that we have, let us agree to colour the point (m, n) in the grid $\{(m, n)\colon m, n \in \mathbb{Z}, m, n \geqslant 0\}$ if we have succeeded in defining $M_{p,q}$ whenever $(w(p), w(q)) = (m, n)$. By (2.52) we can colour $(0, 0)$, $(0, 1)$, $(1, 0)$ and $(1, 1)$. By (2.53a)–(2.55b) we can colour $(m+1, n+1)$ provided (m, n), $(m, n+1)$ and $(m+1, n)$ are coloured. By (2.56a) and (2.57a), if $m \geqslant 1$, $(m+1, 0)$ can be coloured if $(m, 0)$ and $(m, 1)$ are coloured. Similarly, if $(0, m)$ and $(1, m)$ are coloured, $(0, m+1)$ can be coloured. It is now clear how we can colour $\mathbb{Z}_+ \times \{0, 1\}$, namely in the order $(2, 0)$, $(2, 1)$, $(3, 0)$, $(3, 1)$, $(4, 0)$, $(4, 1), \ldots$. It is now simple to complete the colouring of $\mathbb{Z}_+ \times \mathbb{Z}_+$. Thus we have defined $M_{p,q}$ for all pairs p, q of monomials. It is clear that, by induction, $M_{p,q} = M_{q,p}$ is a multiplication operator in the two variables x, y, and that (i) holds.

To prove (iii), by symmetry between the two letters x, y it is enough to prove the formula

$$U_{x^k}M_{p,q} = M_{x^kp,x^kq}.$$

If p, $q \in Y$, this follows from (2.53a). If p, $q \in X$, write $p = x^ir$, $q = x^is$, where $r \in Y$ or $s \in Y$. If the formula we will prove holds for $M_{r,s}$, we get

$$\begin{aligned} U_{x^k}M_{p,q} &= U_{x^k}M_{x^ir,x^is} = U_{x^k}U_{x^i}M_{r,s} = U_{x^{k+i}}M_{r,s} \\ &= M_{x^{k+i}r,x^{k+i}s} = M_{x^kp,x^kq}. \end{aligned}$$

Hence we are reduced to the case where either p or q belongs to Y, but not both. By symmetry we may assume $p \in X_0$, so that $p = x^ir$, $r \in Y$. Since $q \in Y$, we find, using (2.53a),

$$M_{x^kp,x^kq} = M_{x^{k+i}r,x^kq} = U_{x^k}M_{x^ir,q} = U_{x^k}M_{p,q}.$$

This proves (iii).

The main bulk of the proof is to show (iv). The simplest case is $p = q = 1$, in which case (iv) is immediate from (2.52). We shall proceed by induction on the weight, so that we now assume $n \geqslant 1$ and (iv) holds whenever $w(p) + w(q) < n$. We shall prove that it holds when $w(p) + w(q) = n$.

We first assume that p, $q \in X_0$. By the symmetry between p and q we may assume $p = x^ir$, $q = x^js$, where $i \geqslant j \geqslant 1$ and r, $s \in Y$. Using (2.53a) and then (2.48) with $m = j$, $n = m + k$ we have

$$\begin{aligned} U_{x^k,y^l}M_{p,q} &= U_{x^k,y^l}M_{x^ir,x^js} \\ &= U_{x^k,y^l}U_{x^j}M_{x^{i-j}r,s} \\ &= (2U_{x^{j+k},y^l}T_{x^j} - U_{x^{2j+k},y^l})M_{x^{i-j}r,s} \\ &= (2U_{x^{j+k},y^l}U_{x^j,y^0} - U_{x^{2j+k},y^l})M_{x^{i-j}r,s}. \end{aligned}$$

Note that $w(x^{i-j}r) + w(s) < w(p) + w(q)$. Therefore we can use induction on the right-hand side of this formula. We get

$$\begin{aligned} U_{x^k,y^l}M_{p,q} &= U_{x^{j+k},y^l}(M_{x^ir,s} + M_{x^{i-j}r,x^js}) \\ &\quad -\tfrac{1}{2}(M_{x^{i+j+k}r,y^ls} + M_{y^lx^{i-j}r,x^{2j+k}s}) \\ &= U_{x^{j+k},y^l}M_{x^{i-j}r,x^js} - \tfrac{1}{2}M_{y^lx^{i-j}r,x^{2j+k}s} \\ &\quad + U_{x^{j+k},y^l}M_{x^ir,s} - \tfrac{1}{2}M_{x^{i+j+k}r,y^ls}. \end{aligned}$$

Consider the first two terms on the right-hand side. If $i = j$ then $w(x^{i-j}r) + w(x^js) < w(p) + w(q)$, so we can use (iv) with $x^{i-j}r$, x^js replacing p, q by induction. If $i > j$ then $w(x^{i-j}r) + w(x^js) = w(p) + w(q)$, but $(i-j) + j = i < i + j$, so we can use induction on $i + j$ instead. Next consider the last two terms. Here $w(x^ir) + w(s) < w(p) + w(q)$, so we can again use

(iv) by induction. We find then,

$$\begin{aligned} U_{x^k,y^l}M_{p,q} &= \tfrac{1}{2}(M_{x^{i+k}r,y^lx^js} + M_{y^lx^ir,x^{j+k}s}). \\ &= \tfrac{1}{2}(M_{x^kp,y^lq} + M_{y^lp,x^kq}). \end{aligned}$$

This proves (iv) when p, $q \in X_0$.

By the symmetry between the letters x, y we also get (iv) when p, $q \in Y_0$.

It remains to prove (iv) in the case $p \in X$, $q \in Y$; by symmetry, the case $q \in X$, $p \in Y$ will follow. If k, $l \geqslant 1$, (iv) is just a reformulation of (2.55a). (That is, unless $p = q = 1$, but this case has been covered before.) So we must cover the case $k = 0$ or $l = 0$. Again, we may by symmetry assume $l = 0$. In other words the formula we must prove is (remember that $U_{x^k,1} = T_{x^k}$):

$$T_{x^k}M_{p,q} = \tfrac{1}{2}(M_{x^kp,q} + M_{p,x^kq}), \tag{2.58}$$

where $k \geqslant 1$, $p \in X$, $q \in Y$, and $p \neq 1$ or $q \neq 1$. The proof of (2.58) is again broken down into several cases.

First, assume both p and q have weight $\leqslant 1$; that is, $p = x^i$, $q = y^j$. If $i \geqslant k$ then by (2.49)

$$\begin{aligned} T_{x^k}M_{p,q} &= T_{x^k}U_{x^i,y^j} \\ &= \tfrac{1}{2}(U_{x^k}U_{x^{i-k},y^j} + U_{x^{i+k},y^j}) \\ &= \tfrac{1}{2}(M_{x^i,x^ky^j} + M_{x^{i+k},y^j}), \end{aligned}$$

where in the last line we used (2.52) and (2.53a). This is (2.58) for this case. To handle the case $i < k$ we use (2.47) and get

$$\begin{aligned} T_{x^k}M_{p,q} &= T_{x^k}U_{x^i,y^j} \\ &= -T_{x^i}U_{x^k,y^j} + U_{x^i,x^k}T_{y^j} + U_{x^{i+k},y^j}. \end{aligned}$$

By the previous case, (2.58) applies to the first term. For the second term, note that (2.45) implies $U_{x^i,x^k} = U_{x^i}T_{x^{k-i}}$. Also, $T_{y^j} = M_{y^j,1}$, so we can use (2.56a) in the second term and get:

$$\begin{aligned} T_{x^k}M_{p,q} &= -\tfrac{1}{2}(M_{x^{i+k},y^j} + M_{x^k,x^iy^j}) + \tfrac{1}{2}U_{x^i}(M_{x^{k-i}y^j,1} + M_{y^j,x^{k-i}}) + M_{x^{i+k},y^j} \\ &= -\tfrac{1}{2}(M_{x^kp,q} + M_{x^k,x^iy^j}) + \tfrac{1}{2}(M_{x^ky^j,x^i} + M_{x^iy^j,x^k}) + M_{x^kp,q} \\ &= \tfrac{1}{2}(M_{x^kp,q} + M_{p,x^kq}), \end{aligned}$$

where we also used (iii). Thus we have proved (2.58) in the case where both p and q have weight $\leqslant 1$.

We consider next the case when either p or q has weight >1, say $p = x^ir$, $q = y^js$, $r \in Y$, $s \in X$. Again we shall consider the cases $i \geqslant k$ and

$i<k$ separately. First assume $i \geq k$. We start with the following:

$$\begin{aligned} T_{x^k}M_{p,q} &= T_{x^k}M_{x^ir,y^js} \\ &= T_{x^k}(2U_{x^i,y^j}M_{r,s} - M_{y^jr,x^is}). \end{aligned} \tag{2.59}$$

This follows from (2.55a), if $i, j \geq 1$. (Note that either $r \neq 1$ or $s \neq 1$, since p or q has weight >1.) If $j=0$ and $s=1$ it follows from (2.56a). In (2.59) we can apply (2.49) to $T_{x^k}U_{x^i,y^j}$ in the first term and, by induction, (iv) to the second, since $w(y^jr)+w(x^is)<w(p)+w(q)$. The result is

$$T_{x^k}M_{p,q} = (U_{x^k}U_{x^{i-k},y^j} + U_{x^{i+k},y^j})M_{r,s} - \tfrac{1}{2}(M_{x^ky^jr,x^is} + M_{y^jr,x^{i+k}s}).$$

Apply (2.53a) to the third term to extract the factor U_{x^k}, and group with the first term:

$$T_{x^k}M_{p,q} = U_{x^k}(U_{x^{i-k},y^j}M_{r,s} - \tfrac{1}{2}M_{y^jr,x^{i-k}s}) + U_{x^{i+k},y^j}M_{r,s} - \tfrac{1}{2}M_{y^jr,x^{i+k}s}.$$

Since $w(r)+w(s)<w(p)+w(q)$, we can use induction on (iv) and apply (iii) to get

$$\begin{aligned} T_{x^k}M_{p,q} &= \tfrac{1}{2}U_{x^k}M_{x^{i-k}r,y^js} + \tfrac{1}{2}M_{x^{i+k}r,y^js} \\ &= \tfrac{1}{2}(M_{x^ir,x^ky^js} + M_{x^{i+k}r,y^js}), \end{aligned}$$

which is (2.58) again.

We are left with the case $i<k$. Equation (2.59) is still valid. We should note that now $i=0$, $r=1$ is possible, but (2.59) still follows from (2.57b) in this case. We now apply (2.47) to $T_{x^k}U_{x^i,y^j}$ in (2.59), and use (iv), by induction, on $T_{x^k}M_{y^jr,x^is}$:

$$\begin{aligned} T_{x^k}M_{p,q} = {} & 2(-T_{x^i}U_{x^k,y^j} + U_{x^i,x^k}T_{y^j} + U_{x^{i+k},y^j})M_{r,s} \\ & - \tfrac{1}{2}(M_{x^ky^jr,x^is} + M_{y^jr,x^{k+i}s}). \end{aligned}$$

In the following calculation, we use (2.49) in the first term above, apply (iii) and, by induction, (iv):

$$\begin{aligned} T_{x^k}M_{p,q} &= -(U_{x^i}U_{x^{k-i},y^j} + U_{x^{i+k},y^j})M_{r,s} + U_{x^i,x^k}(M_{y^jr,s} + M_{r,y^js}) \\ &\quad + (M_{x^{i+k}r,y^js} + M_{y^jr,x^{i+k}s}) - \tfrac{1}{2}(M_{x^ky^jr,x^is} + M_{y^jr,x^{i+k}s}) \\ &= -\tfrac{1}{2}U_{x^i}(M_{x^{k-i}r,y^js} + M_{y^jr,x^{k-i}s}) - \tfrac{1}{2}(M_{x^{i+k}r,y^js} + M_{y^jr,x^{i+k}s}) \\ &\quad + \tfrac{1}{2}(M_{x^iy^jr,x^ks} + M_{x^ky^jr,x^is} + M_{x^ir,x^ky^js} + M_{x^kr,x^iy^js}) \\ &\quad + M_{x^{i+k}r,y^js} + \tfrac{1}{2}M_{y^jr,x^{i+k}s} - \tfrac{1}{2}M_{x^ky^jr,x^is} \\ &= -\tfrac{1}{2}M_{x^kr,x^iy^js} - \tfrac{1}{2}M_{x^iy^jr,x^ks} + \tfrac{1}{2}M_{x^iy^jr,x^ks} \\ &\quad + \tfrac{1}{2}M_{x^ir,x^ky^js} + \tfrac{1}{2}M_{x^kr,x^iy^js} + \tfrac{1}{2}M_{x^{i+k}r,y^js} \\ &= \tfrac{1}{2}(M_{x^ir,x^ky^js} + M_{x^{i+k}r,y^js}), \end{aligned}$$

which is again (2.58). This finishes the proof!

2.4.25. Proof of Macdonald's theorem (2.4.13). We adopt the notation of 2.4.24. We get a linear map $t \to M_t$ from the tensor product

$$FA\{x, y\} \underset{\Phi}{\otimes} FA\{x, y\}$$

into the algebra of multiplication operators on $FJ\{x, y, z\}$, defined by $M_{p \otimes q} = M_{p,q}$. Since M_t only depends on the symmetric part of t, we may restrict the map to the space S of symmetric tensors. Let E be the range of the map. By 2.4.24(ii), (iii) and (iv), $1 \in E$, and E is invariant under left multiplication with $\{U_{a,b}: a, b \in \{1, x, y\}\}$. By 2.4.23 E then contains the algebra of multiplication operators in x, y, so we have: every multiplication operator in x, y on $FJ\{x, y, z\}$ is of the form M_t for some $t \in S$.

Assume now that $t \in S$, and M_t vanishes on $FS\{x, y, z\}$. To finish the proof, we must show it vanishes on $FJ\{x, y, z\}$. More precisely, we shall use that $M_t(z) = 0$ in $FS\{x, y, z\}$ to conclude that $t = 0$.

From 2.4.24(i) we see that what must be proved is the following. Consider the map γ: $FA\{x, y\} \otimes FA\{x, y\} \to FS\{x, y, z\}$ defined by $\gamma(p \otimes q) = \frac{1}{2}(pzq^* + qzp^*)$. Then the restriction of γ to the space S of symmetric tensors is injective.

To prove this, note that the monomials form a basis for $FA\{x, y\}$, so that the family of $\frac{1}{2}(p \otimes q + q \otimes p)$, indexed by the unordered pair $\{p, q\}$, is a basis for S. Therefore it is enough to show that the image of this basis, $\frac{1}{2}(pzq^* + qzp^*)$, indexed by the unordered pair $\{p, q\}$, is linearly independent. But pzq^* and qzp^* are monomials in $FA\{x, y, z\}$, and the monomials in $FA\{x, y, z\}$ are linearly independent. If p, q, r, s are monomials in $FA\{x, y\}$ and $\{pzq^*, qzp^*\} \cap \{rzs^*, szr^*\} \neq \varnothing$, then $\{p, q\} = \{r, s\}$ (note the role of z as a 'separator' here). The wanted linear independence follows easily.

2.5. Operator commutation and the centre

2.5.1. Consider a pair a, b in an associative algebra A. We may ask: what properties of the pair a, b can be expressed, in terms of Jordan structure, corresponding to the commutation relation $ab = ba$? Of course, the Jordan algebra A^J is commutative. However, it need not be associative, and this suggests two conditions: for any $c \in A^J$, $a \circ (c \circ b) = (a \circ c) \circ b$, or the subalgebra of A^J generated by a, b is associative. It is clear that these two conditions are implied by the relation $ab = ba$.

In general two elements a, b in a Jordan algebra A are said to *operator commute* if the operators T_a, T_b commute, i.e. if $(a \circ c) \circ b = a \circ (c \circ b)$ for all $c \in A$. We shall see later that, if A is a JB algebra, this is equivalent to stating that a and b generate an associative subalgebra.

By the *centre* of A we mean the set of all elements of A which operator commute with every other element of A.

2.5.2. We now give two examples to show that in general the two conditions $T_aT_b = T_bT_a$ and 'a, b generate an associative Jordan algebra' are independent.

First, let A be the algebra of linear maps of the polynomial ring $\mathbb{R}[x]$ into itself. Let p be derivation and q multiplication by x. Then $[p, q] = 1$, since $(xf)' = f + xf'$. For any $r \in A$ we then have

$$[T_p, T_q]r = p\circ(q\circ r) - q\circ(p\circ r) = \tfrac{1}{4}[[p, q], r] = 0,$$

so that p and q operator commute. However, we find $pq^2 = 2q + q^2p$, or $[p, q^2] = 2q$. We then get

$$[T_p, T_{q^2}]p = \tfrac{1}{4}[[p, q^2], p] = -\tfrac{1}{2},$$

so $(q^2\circ p)\circ p \neq q^2\circ(p\circ p)$. Therefore, p, q do not generate an associative subalgebra of A^J.

For an example in the converse direction let A be the 3×3 matrix algebra over the reals, and let

$$e_{12} = \begin{pmatrix} 0 & 1 & 0 \\ 0 & 0 & 0 \\ 0 & 0 & 0 \end{pmatrix}, \quad e_{23} = \begin{pmatrix} 0 & 0 & 0 \\ 0 & 0 & 1 \\ 0 & 0 & 0 \end{pmatrix}, \quad e_{13} = \begin{pmatrix} 0 & 0 & 1 \\ 0 & 0 & 0 \\ 0 & 0 & 0 \end{pmatrix}.$$

Then $e_{12}\circ e_{23} = \frac{1}{2}e_{13}$, while $e_{12}\circ e_{13} = e_{23}\circ e_{13} = e_{13}^2 = e_{12}^2 = e_{23}^2 = 0$. Thus e_{12} and e_{23} generate an associative subalgebra of A^J, but they do not operator commute since $[T_{e_{12}}, T_{e_{23}}]e_{11} = -\frac{1}{4}e_{13} \neq 0$.

2.5.3. Lemma. *The centre of a Jordan algebra is an associative subalgebra.*

Proof. The centre is clearly a linear subspace. If x, y belong to the centre and a, b are in the given algebra, we have

$$\begin{aligned} T_{x\circ y}T_a b &= (x\circ y)\circ(a\circ b) = x\circ(y\circ(a\circ b)) \\ &= x\circ((y\circ b)\circ a) = (x\circ(y\circ b))\circ a \\ &= ((x\circ y)\circ b)\circ a = T_a T_{x\circ y} b, \end{aligned}$$

which proves that $x\circ y$ belongs to the centre. The associativity of the centre is immediate.

2.5.4. Let p be an idempotent in a unital Jordan algebra A, i.e. $p^2 = p$. Let $p^\perp = 1 - p$. Then a straightforward calculation shows that

$$p\circ a = \tfrac{1}{2}(a + \{pap\} - \{p^\perp a p^\perp\}),$$

or

$$T_p = \tfrac{1}{2}(\iota + U_p - U_{p^\perp}). \tag{2.60}$$

From Macdonald's theorem we have for any $a \in A$ that $U_a^2 = U_{a^2}$ and $U_a U_{1-a} = U_{a-a^2}$. Applied to $a = p$ this becomes

$$U_p^2 = U_p; \qquad U_{p^\perp}^2 = U_{p^\perp}; \qquad U_p U_{p^\perp} = U_{p^\perp} U_p = 0. \tag{2.61}$$

Thus, (2.60) can be considered a 'spectral decomposition' for T_p. Combining (2.60) and (2.61) we get in particular,

$$T_p U_p = U_p T_p = U_p; \qquad T_p U_{p^\perp} = U_{p^\perp} T_p = 0. \tag{2.62}$$

2.5.5. Lemma. *Let A be a unital Jordan algebra and p an idempotent in A. For any $a \in A$ the following conditions are equivalent:*

(i) *a and p operator commute,*
(ii) *$T_p a = U_p a$,*
(iii) *$a = (U_p + U_{p^\perp})a$,*
(iv) *a and p generate an associative subalgebra of A.*

Moreover, U_pA and $U_{p^\perp}A$ are subalgebras of A, and $a \circ b = 0$ if $a \in U_pA$, $b \in U_{p^\perp}A$.

Proof. (i) $\Rightarrow$ (ii) From (i) we get $0 = [T_p, T_a]p = p \circ (a \circ p) - p \circ a$, or $T_p a = T_p^2 a$. But then $U_p a = (2T_p^2 - T_p)a = T_p a$, which is (ii).

(ii) $\Rightarrow$ (iii) is immediate from (2.60).

(iii) $\Rightarrow$ (i) From the linearized Jordan axiom (2.33) with $b = c = p$ we get

$$2[T_p, T_{p\circ a}] = [T_p, T_a]. \tag{2.63}$$

Let $r = U_p a$, $s = U_{p^\perp} a$. By (2.62), $p \circ s = 0$, so (2.63) implies $[T_p, T_s] = 0$. Also by (2.62), $p \circ r = r$, so again, by (2.63), $[T_p, T_r] = 0$. Hence $a = r + s$ operator commutes with p.

Before showing the equivalence of (iv) with the others, we prove the final statement of the lemma. From 2.4.17 we have $\{pap\}^2 = \{p\{ap^2a\}p\}$ for any $a \in A$. This shows that U_pA is closed under squaring and is therefore a subalgebra of A. Similarly $U_{p^\perp}A$ is a subalgebra of A. Let $a \in U_pA$, $b \in U_{p^\perp}A$. Since (iii) $\Rightarrow$ (i), a operator commutes with p. This implies $T_{p^\perp}(a \circ b) = a \circ b$, and hence $U_{p^\perp}(a \circ b) = (2T_{p^\perp}^2 - T_{p^\perp})(a \circ b) = a \circ b$. Interchanging a, b and also p, $p^\perp$, we similarly get $U_p(a \circ b) = a \circ b$. Hence $a \circ b = U_p U_{p^\perp}(a \circ b) = 0$, by (2.61).

(iii) $\Rightarrow$ (iv) Assume (iii) holds. Let $b = U_p a$, $c = U_{p^\perp} a$. The algebra B generated by a, p and 1 is then generated by b, p, c and $p^\perp$. But $b \in U_pA$ and p is by (2.62) the unit of U_pA. By power associativity (2.4.4) b and p generate an associative subalgebra B_1 of U_pA. Similarly, c and $p^\perp$

generate an associative subalgebra B_2 of $U_{p^\perp}A$. Hence $B=B_1\oplus B_2$ is associative, and (iv) follows.

(iv) $\Rightarrow$ (ii) If (iv) holds then $p\circ(p\circ a)=p^2\circ a=p\circ a$, i.e. $T_p^2a=T_pa$. Thus $U_pa=(2T_p^2-T_p)a=T_pa$.

2.5.6. Lemma. *Let e be an idempotent in a unital Jordan algebra. Then e belongs to the centre if and only if T_e is a homomorphism.*

Proof. If e belongs to the centre then by 2.5.5 $A=U_eA\oplus U_{e^\perp}A$, and moreover T_e is the projection on the first summand. Clearly then T_e is a homomorphism.

Conversely, if T_e is a homomorphism and a is arbitrary, then $T_ea=T_e(a\circ 1)=(T_ea)\circ e=T_e^2a$, so $T_e^2=T_e$. Since $U_e=(2T_e^2-T_e)=T_e$, e is central by 2.5.5.

2.5.7. From the above proof it is clear that, if e is a central idempotent in a unital Jordan algebra A, then U_eA is a direct summand and, in particular, an ideal in A. Conversely, if e is an idempotent in A such that U_eA is an ideal, then e is central. For then, if $a\in A$, we must have $e\circ a\in U_eA$, i.e. $U_e(e\circ a)=e\circ a$. By (2.62), however, this implies $U_ea=T_ea$. Hence by 2.5.5, e is central.

2.5.8. Central projections can be used to construct more general ideals. For example, if A is a unital Jordan algebra, B a subalgebra of A and e a central projection in A (or more generally e operator commutes with all $b\in B$), then the set of all $b\in B$ such that $e\circ b=0$ is an ideal in B. For if $e\circ b=0$ and $c\in B$, then $e\circ(b\circ c)=c\circ(e\circ b)=0$.

2.6. Peirce decomposition

2.6.1. Consider a unital, associative algebra $\mathcal{A}$. If $\mathcal{A}$ contains orthogonal idempotents $p_1,\ldots,p_n$ (i.e. $p_i^2=p_i$, $p_ip_j=p_jp_i=0$ if $i\neq j$) with sum 1, we have a decomposition $\mathcal{A}=\bigoplus_{i,j}\mathcal{A}_{ij}$, where $\mathcal{A}_{ij}=p_i\mathcal{A}p_j$. We may think of $\mathcal{A}$ as consisting of *matrices* $a=(a_{ij})$, where $a_{ij}\in\mathcal{A}_{ij}$. The product in $\mathcal{A}$ is like matrix multiplication in that $ab=(\sum_k a_{ik}b_{kj})$ if $a=(a_{ij})$, $b=(b_{ij})$. The present section is devoted to the generalization of this idea to Jordan algebras. Of course, $p_i\mathcal{A}p_j$ does not make sense in the Jordan algebra setting, but $p_i\mathcal{A}p_j+p_j\mathcal{A}p_i=\{p_i\mathcal{A}p_j\}$ does. The analogy with matrices should be kept in mind when reading this section.

2.6.2. Throughout Section 2.6, A will be a unital Jordan algebra. Let p be an idempotent in A. Recall that by (2.60) we have

$$T_p=\tfrac{1}{2}(\iota+U_p-U_{p^\perp}).$$

Also, by (2.61), U_p and $U_{p^\perp}$ are orthogonal, idempotent mappings. Rewriting the above formula as

$$T_p = U_p + \tfrac{1}{2}(\iota - U_p - U_{p^\perp}) + 0\cdot U_{p^\perp}, \tag{2.64}$$

and noting that U_p, $\iota - U_p - U_{p^\perp}$ and $U_{p^\perp}$ are mutually orthogonal idempotent mappings with sum 1, we must conclude that T_p has eigenvalues 1, $\frac{1}{2}$ and 0, and that we have the following vector space decomposition:

$$A = A_1 \oplus A_{1/2} \oplus A_0, \tag{2.65}$$

where A_i is the eigenspace corresponding to the eigenvalue i ($i = 0, \frac{1}{2}, 1$). This is called the *Peirce decomposition* of A with respect to p.

From the definition of the triple product mapping we get

$$\begin{aligned} U_{p,p^\perp} &= T_pT_{p^\perp} + T_{p^\perp}T_p - T_{p\circ p^\perp} \\ &= T_p(\iota - T_p) + (\iota - T_p)T_p - 0 = 2(T_p - T_p^2). \end{aligned}$$

From (2.64) we then get

$$U_{p,p^\perp} = \tfrac{1}{2}(\iota - U_p - U_{p^\perp}).$$

Thus U_p, $2U_{p,p^\perp}$ and $U_{p^\perp}$ are the projections on the direct summands A_1, $A_{1/2}$ and A_0 respectively corresponding to the Peirce decomposition.

2.6.3. Lemma. *Let A be a unital Jordan algebra and p an idempotent in A. Let $A = A_1 \oplus A_{1/2} \oplus A_0$ be the corresponding Peirce decomposition. Then we have the following multiplication rules:*

$$A_0 \circ A_0 \subseteq A_0; \qquad A_1 \circ A_1 \subseteq A_1; \qquad A_0 \circ A_1 = 0;$$
$$(A_0 \oplus A_1) \circ A_{1/2} \subseteq A_{1/2}; \qquad A_{1/2} \circ A_{1/2} \subseteq A_0 \oplus A_1.$$

If $a \in A_0$, $b \in A_1$, then a and b operator commute.

Proof. The first three rules follow from 2.5.5. If $a \in A_0 \oplus A_1$ then by 2.5.5 a operator commutes with p. Therefore, if $b \in A_{1/2}$, $p \circ (a \circ b) = a \circ (p \circ b) = \frac{1}{2} a \circ b$, so that $a \circ b \in A_{1/2}$. Next, assume a, $b \in A_{1/2}$. The linearized Jordan axiom yields

$$\begin{aligned} 0 &= [T_a, T_{p\circ b}] + [T_b, T_{a\circ p}] + [T_p, T_{a\circ b}] \\ &= \tfrac{1}{2}[T_a, T_b] + \tfrac{1}{2}[T_b, T_a] + [T_p, T_{a\circ b}] \\ &= [T_p, T_{a\circ b}]. \end{aligned}$$

By 2.5.5 then $a \circ b \in A_0 \oplus A_1$. This proves the multiplication rules. If $a \in A_0$, $b \in A_1$ then $p \circ b = b$, $p \circ a = 0$, $a \circ b = 0$, so the linearized Jordan axiom, as above, yields $[T_a, T_b] = 0$.

2.6.4. Two idempotents p, q will be called *orthogonal* if $p\circ q=0$. We shall generalize the Peirce decomposition to the case of several orthogonal idempotents. Thus the decomposition described in 2.6.5 will be called the Peirce decomposition with respect to $p_1, \ldots, p_n$.

2.6.5. Theorem. *Let A be a unital Jordan algebra. Suppose $p_1, \ldots, p_n$ are pairwise orthogonal idempotents in A with sum 1. Let $A_{ij}=\{p_iAp_j\}$. Then $A_{ij}=A_{ji}$, and we have the decomposition*

$$A=\bigoplus_{1\leqslant i\leqslant j\leqslant n} A_{ij}.$$

Furthermore, the following multiplication table holds:

$$A_{ij}\circ A_{kl}=0 \qquad \text{if } \{i,j\}\cap\{k,l\}=\varnothing, \tag{2.66}$$

$$A_{ij}\circ A_{kj}\subseteq A_{ik} \qquad \text{if } i, j, k \text{ are all distinct}, \tag{2.67}$$

$$A_{ij}\circ A_{ij}\subseteq A_{ii}+A_{jj}, \tag{2.68}$$

$$A_{ii}\circ A_{ij}\subseteq A_{ij}. \tag{2.69}$$

Proof. That $A_{ij}=A_{ji}$ is trivial. It is also clear that $A=\sum A_{ij}$, since if $a\in A$ we can write $a=\{1a1\}=\sum\{p_iap_j\}$. To show that the sum is direct, it will be enough to show

$$T_{p_k}\mid A_{ij}=\begin{cases} 0 & \text{if } k\notin\{i,j\},\\ 1 & \text{if } i=j=k,\\ \tfrac12 & \text{if } i\neq j \text{ and } k\in\{i,j\}.\end{cases} \tag{2.70}$$

Let $A=A_1^k\oplus A_{1/2}^k\oplus A_0^k$ be the Peirce decomposition with respect to p_k. Then

$$A_1^k=\{p_kAp_k\}=A_{kk},$$

$$A_{1/2}^k=\{p_kAp_k^\perp\}=\sum_{i\neq k}\{p_kAp_i\}=\sum_{i\neq k}A_{ki},$$

$$A_0^k=\{p_k^\perp Ap_k^\perp\}=\sum_{i\neq k}\sum_{j\neq k}\{p_iAp_j\}=\sum_{i\neq k}\sum_{j\neq k}A_{ij},$$

and from these the assertion follows.

The multiplication rules (2.66)–(2.69) are more or less trivial consequences of 2.6.3 together with (2.70). Indeed, to show (2.66), let $p=p_i+p_j$ if $i\neq j$, $p=p_i$ otherwise. Then, by (2.70) $A_{ij}\subseteq A_1$ and $A_{kl}\subset A_0$, where $A=A_1\oplus A_{1/2}\oplus A_0$ is the Peirce decomposition relative to p. Then (2.66) follows from 2.6.3. To show (2.67), use the Peirce decomposition with respect to k. Since $A_{ij}\subseteq A_0^k$, $A_{kj}\subseteq A_{1/2}^k$, 2.6.3 yields $A_{ij}\circ A_{kj}\subseteq A_{1/2}^k$. Similarly $A_{ij}\circ A_{kj}\subseteq A_{1/2}^i$, so $A_{ij}\circ A_{kj}\subseteq A_{1/2}^k\cap A_{1/2}^i=A_{ik}$, by the above formulae. Relations (2.68) and (2.69) in the case $i=j$, follow directly from 2.6.3 with $p=p_i$. If $i\neq j$ they also follow from 2.6.3 with $p=p_i$, and working in the subalgebra $U_{p_i+p_j}(A)$.

2.6.6. Lemma. *Let A be a unital Jordan algebra and $p_1,\ldots,p_n$ pairwise orthogonal idempotents in A with sum 1. Let $A_{ij}=\{p_iAp_j\}$. Then:*
(i) $\{A_{ij}A_{jk}A_{ki}\}\subseteq A_{ii}$ *for all i, j, k.*
(ii) $\{A_{ij}A_{jk}A_{ij}\}=0$ *if i, j, k are distinct.*

Proof. (i) Let $a\in A_{ij}$, $b\in A_{jk}$, $c\in A_{ki}$. From the definition of the triple product and 2.6.5 we have

$$\{abc\}\in A_{ii}+A_{jj}+A_{kk}.$$

We show that, if $j\neq i$ the A_{jj} component is zero. The same result for the A_{kk} component is similar. From 2.4.20 we have

$$p_j\circ\{abc\}=\{(p_j\circ a)bc\}+\{ab(p_j\circ c)\}-\{a(p_j\circ b)c\}.$$

Then, from (2.70) we get $p_j\circ\{abc\}=0$ whether $k=i$, $k=j$, or $i\neq k\neq j$. This proves (i).

To show (ii) suppose $a\in A_{ij}$, $b\in A_{jk}$, $c\in A_{ij}$, and let $p=p_i+p_j$. Then $a=\{pap\}$, so from 2.4.18 we get

$$\{aba\}=\{\{pap\}b\{pap\}\}=\{p\{a\{pbp\}a\}p\}=0,$$

since $\{pbp\}=0$. Similarly, $\{cbc\}=0$, and $\{(a+c)b(a+c)\}=0$. Combining these identities, we get $\{abc\}=0$.

2.7. Jordan matrix algebras

2.7.1. Let R be any algebra. Then $M_n(R)$, the space of $n\times n$ matrices with coefficients in R, is also an algebra with the usual matrix product: $(a_{ij})(b_{jk})=(\sum_j a_{ij}b_{jk})$. If R is, moreover, a $*$ algebra then so is $M_n(R)$ with $(a_{ij})^*=(a_{ji}^*)$. The Hermitian, or self-adjoint, part of $M_n(R)$ is denoted by $H_n(R)$. $H_n(R)$ is also an algebra, with the product defined by $a\circ b=\frac{1}{2}(ab+ba)$. $H_n(R)$ may be a Jordan algebra. This is certainly the case if R is associative, but, as will be seen below, that condition is not necessary. A Jordan algebra of the form $H_n(R)$ will be called a *Jordan matrix algebra.*

Jordan matrix algebras of prime concern to us will be $H_n(\mathbb{R})$, $H_n(\mathbb{C})$, $H_n(\mathbb{H})$ and $H_3(\mathbb{O})$. Since $\mathbb{R}$, $\mathbb{C}$ and $\mathbb{H}$ are associative it is clear that the first three are Jordan matrix algebras. Our first objective will be to prove the corresponding statement for $H_3(\mathbb{O})$. Then we proceed to show that, over the reals and with mild restrictions, the above-mentioned examples are canonical.

We are really only interested in Jordan matrix algebras $H_n(R)$ for

$n \geq 3$. The reason for this is that then the product in R can be reconstructed from the Jordan product in $H_n(R)$ using formulae like

$$\begin{pmatrix} 0 & a & 0 \\ a^* & 0 & 0 \\ 0 & 0 & 0 \end{pmatrix} \circ \begin{pmatrix} 0 & 0 & 0 \\ 0 & 0 & b \\ 0 & b^* & 0 \end{pmatrix} = \tfrac{1}{2} \begin{pmatrix} 0 & 0 & ab \\ 0 & 0 & 0 \\ b^*a^* & 0 & 0 \end{pmatrix}.$$

In the case $n=2$ we get algebras looking like the spin factors studied in Chapter 6 (see also 2.9.7).

2.7.2. Since we shall work quite a bit with matrix algebras it is useful to introduce some compact notation. Let R be a unital $*$ algebra. The *matrix units* in $M_n(R)$ are the elements e_{ij}, where $1 \leq i \leq n$ and $1 \leq j \leq n$. Here, e_{ij} is the matrix whose (i, j) entry is 1, the others being zero. We then write ae_{ij} for the matrix whose (i, j) entry is a, the others being zero. This is consistent with identifying $a \in R$ with $\sum_{i=1}^n ae_{ii} \in M_n(R)$. The matrix (a_{ij}) may then be written as $\sum_{ij} a_{ij}e_{ij}$. Note that $e_{ij}^* = e_{ji}$, $\sum_{i=1}^n e_{ii} = 1$, the identity in $M_n(R)$, and that $e_{ij}e_{kl} = 0$ if $j \neq k$, $e_{ij}e_{jk} = e_{ik}$.

2.7.3. Lemma. *Let $a, b, c, d \in \mathbb{O}$. Then we have:*
(i) $[d, [a, b, c]] + [ab, c, d] + [bc, a, d] + [ca, b, d] = 0$,
(ii) $\overline{[a, b, c]} = -[\bar{c}, \bar{b}, \bar{a}] = -[a, b, c]$,
(iii) $[a, b, c] = -[\bar{a}, b, c] = -[a, \bar{b}, c] = -[a, b, \bar{c}]$.

Proof. Recall the definition of the associator, in any algebra:

$$[a, b, c] = (ab)c - a(bc).$$

Let us linearize the Moufang identity (2.2.2) $(ca)(bc) = (c(ab))c$, i.e. substitute $c+d$ for c, and subtract the original identity twice, the second time with d replacing c. We get

$$(ca)(bd) + (da)(bc) = (c(ab))d + (d(ab))c.$$

In the following calculation we use this formula and the fact that associators alternate in $\mathbb{O}$:

$$\begin{aligned} d[a, b, c] &= d((ab)c - a(bc)) \\ &= -[d, ab, c] + (d(ab))c - d(a(bc)) \\ &= -[ab, c, d] + (d(ab))c + [d, a, bc] - (da)(bc) \\ &= -[ab, c, d] - [bc, a, d] - (c(ab))d + (ca)(bd) \\ &= -[ab, c, d] - [bc, a, d] + [c, a, b]d - ((ca)b)d \\ &\quad + (ca)(bd) \\ &= -[ab, c, d] - [bc, a, d] + [c, a, b]d - [ca, b, d], \end{aligned}$$

which is (i). The first half of (ii) follows from $\overline{ab}=\bar{b}\bar{a}$. The second half of (ii) follows from (iii) and the fact that associators alternate. Finally, (iii) follows from the fact that $a+\bar{a}$ is a real scalar, so that $[a+\bar{a}, b, c]=0$ (and similarly in the other variables).

2.7.4. Lemma. *Let* $x=(a_{ij})\in H_3(\mathbb{O})$. *Then* $[x^2, x]=2[a_{12}, a_{23}, a_{31}]\cdot 1$.

Proof. We get

$$[x^2, x]=[x, x, x]=\sum_{i,j,k,l}[a_{ij}, a_{jk}, a_{kl}]e_{il}. \tag{2.71}$$

If $i=j$ then $a_{ij}\in\mathbb{R}$, so $[a_{ij}, a_{jk}, a_{kl}]=0$. Similarly if $j=k$ or $k=l$. Hence, the only nonzero terms in the above sum are those for which $i\neq j\neq k\neq l$. Furthermore, if $i=k$ then

$$[a_{ij}, a_{jk}, a_{kl}]=[a_{ij}, a_{ji}, a_{kl}]=[a_{ij}, \bar{a}_{ij}, a_{kl}]=-[a_{ij}, a_{ij}, a_{kl}]=0$$

by 2.7.3(iii) and the alternative law. Similarly, if $j=l$ then $[a_{ij}, a_{jk}, a_{kl}]=0$. It follows therefore that the only nonzero terms on the right-hand side of (2.71) are those for which i, j, k are distinct, and $l=i$ (since there are only three distinct possibilities for the indices). Therefore (2.71) becomes the following, where the sum is over indices such that i, j, k are distinct:

$$[x^2, x]=\sum [a_{ij}, a_{jk}, a_{ki}]e_{ii}. \tag{2.72}$$

Using 2.7.3(iii), the fact that $\bar{a}_{ij}=a_{ji}$ and the fact that associators alternate we see that $[a_{ij}, a_{jk}, a_{ki}]$ is invariant under the transposition of any two indices, and therefore under any permutation. Thus (2.72) implies

$$[x^2, x]=2[a_{12}, a_{23}, a_{31}]\sum e_{ii},$$

which completes the proof.

2.7.5. Theorem. $H_3(\mathbb{O})$ *is a Jordan algebra.*

Proof. Since the product defined by $x\circ y=\frac{1}{2}(xy+yx)$ is clearly commutative, we must only verify the identity $(x^2\circ y)\circ x=x^2\circ(y\circ x)$, for $x, y\in H_3(\mathbb{O})$. Let us write $[x, y, z]^J$ for the associator relative to this product, i.e.

$$[x, y, z]^J=(x\circ y)\circ z-x\circ(y\circ z).$$

By symmetry in the indices, it is enough to show that the (1, 1) and (1, 2) entries of $[x, y, x^2]^J$ vanish.

We set

$$x=\sum a_{ij}e_{ij}, \qquad y=\sum b_{ij}e_{ij},$$

$$x^2=\sum c_{ij}e_{ij}, \qquad c_{ij}=\sum a_{ik}a_{kj}=\sum a_{ik}\bar{a}_{jk}.$$

The following identity is immediate, where $[a, b, c]$ denotes the associator as computed in the ordinary matrix product on $M_3(\mathbb{O})$, i.e. $[a, b, c]=(ab)c-a(bc)$:

$$\begin{aligned}4[x, y, x^2]^J=&[x, y, x^2]-[x^2, y, x]+[y, x, x^2]-[x^2, x, y]\\&+[x, x^2, y]-[y, x^2, x]+[y, [x, x^2]]. \qquad (2.73)\end{aligned}$$

By 2.7.4, $[x, x^2]=-2a\mathbf{1}$, where $a=[a_{12}, a_{23}, a_{31}]$. So the last term in (2.73) is

$$[y, [x, x^2]]=-2[y, a\mathbf{1}]=2\sum [a, b_{ij}]e_{ij}. \qquad (2.74)$$

Each matrix entry of an associator $[u, v, w]$ where $u, v, w\in H_3(\mathbb{O})$ is itself a sum of associators in $\mathbb{O}$, and is therefore skew by 2.7.3(ii). From this fact, together with (2.73) and (2.74), we find that the (1, 1) entry of $[x, y, x^2]^J$ is skew. But because $[x, y, x^2]^J\in H_3(\mathbb{O})$, the same entry must be Hermitian, and then it must be zero.

It remains to show that the (1, 2) entry of $[x, y, x^2]^J$ vanishes. By linearity in y it is enough to check this in the following cases:

(a) $y=be_{ii}$, $(b=\bar{b})$,

(b) $y=be_{13}+\bar{b}e_{31}$,

(c) $y=be_{12}+\bar{b}e_{21}$.

The case $y=be_{23}+\bar{b}e_{32}$ follows from (b) by the symmetry in the indices.

(a) If $y=be_{ii}$, $b\in\mathbb{R}$ so all the associators on the right-hand side of (2.73) will vanish, since any associator in $\mathbb{O}$ involving b must vanish. By (2.74), the last term in (2.73) also vanishes in this case, so $[x, y, x^2]^J=0$.

(b) In the case $y=be_{13}+\bar{b}e_{31}$ we find that the (1, 2) entry of $[x, y, x^2]$ is

$$\sum_{k,l} [a_{1k}, b_{kl}, c_{l2}]=[a_{11}, b, c_{32}]+[a_{13}, \bar{b}, c_{12}]=[a_{13}, \bar{b}, c_{12}]$$

because $a_{11}\in\mathbb{R}$. We find similar formulae for the (1, 2) entries of the other terms on the right-hand side of (2.73), using (2.74) to see that $[y, [x, x^2]]$ has no (1, 2) term. We conclude that the (1, 2) term of $4[x, y, x^2]^J$ is

$$[a_{13}, \bar{b}, c_{12}]-[c_{13}, \bar{b}, a_{12}]+[b, \bar{a}_{13}, c_{12}]-0+0-[b, \bar{c}_{13}, a_{12}]+0,$$

which vanishes by 2.7.3(iii) and the fact that associators in $\mathbb{O}$ alternate.

(c) If $y = be_{12} + \bar{b}e_{21}$, we find by a similar argument, using (2.73) and (2.74), that the (1, 2) term of $4[x, y, x^2]^J$ is

$$\begin{aligned}2d = &[a_{12}, \bar{b}, c_{12}] - [c_{12}, \bar{b}, a_{12}] + [b, a_{21}, c_{12}]\\ &+ [b, a_{23}, c_{32}] - [c_{12}, a_{21}, b] - [c_{13}, a_{31}, b]\\ &+ [a_{12}, c_{21}, b] + [a_{13}, c_{31}, b] - [b, c_{21}, a_{12}]\\ &- [b, c_{23}, a_{32}] - 2[b, a],\end{aligned}$$

where $a = [a_{12}, a_{23}, a_{31}]$. Again, using 2.7.3(iii) and the fact that associators in $\mathbb{O}$ alternate, we can reduce this to

$$-d = [c_{12}, a_{21}, b] + [c_{23}, a_{32}, b] + [c_{13}, a_{31}, b] + [b, a]. \tag{2.75}$$

Now we use $c_{ij} = \sum a_{ik}a_{kj}$ and the fact that $a_{ii}, a_{jj} \in \mathbb{R}$ to get

$$[c_{ij}, a_{ji}, b] = (a_{ii} + a_{jj})[a_{ij}, a_{ji}, b] + [a_{ik}a_{kj}, a_{ji}, b]$$

where i, j, k are distinct. Using 2.7.3(iii) and the alternative law we find $[a_{ij}, a_{ji}, b] = [a_{ij}, \bar{a}_{ij}, b] = -[a_{ij}, a_{ij}, b] = 0$, so the above reduces to

$$[c_{ij}, a_{ji}, b] = [a_{ik}a_{kj}, a_{ji}, b]$$

with i, j, k distinct. Substitute this into (2.75) to get

$$\begin{aligned}-d &= [a_{13}a_{32}, a_{21}, b] + [a_{21}a_{13}, a_{32}, b] + [a_{12}a_{23}, a_{31}, b] + [b, a]\\ &= [a_{23}a_{31}, a_{12}, b] + [a_{31}a_{12}, a_{23}, b]\\ &\quad + [a_{12}a_{23}, a_{31}, b] + [b, [a_{12}, a_{23}, a_{31}]] = 0,\end{aligned}$$

where we first used 2.7.3(iii) on the first two terms and then 2.7.3(i) to see that it all vanishes. But this ends the proof.

2.7.6. Theorem. *Let $A = H_n(R)$ be a Jordan matrix algebra. Then we have:*
(i) *if $n \geq 4$ then R is associative,*
(ii) *if $n \geq 3$ then R is clternative.*

Proof. (i) Let $x, y, z \in R$ and define four elements of A as follows: $a = xe_{12} + x^*e_{21}$, $b = ye_{23} + y^*e_{32}$, $c = ze_{34} + z^*e_{43}$, $p = e_{11} + e_{22}$. Then p is an idempotent, giving rise to a Peirce decomposition $A = A_1 \oplus A_{1/2} \oplus A_0$. Clearly, $a \in A_1$ and $c \in A_0$. Therefore, by 2.6.3, a and c operator commute, so that $(a \circ b) \circ c = a \circ (b \circ c)$. Multiplying out the matrices, we get $(xy)z = x(yz)$.

(ii) Let $n = 3$. If $x = x^*$, $y, z \in R$, we can mimick the above proof, setting $a = xe_{11}$, $b = ye_{12} + y^*e_{21}$, $c = ze_{23} + z^*e_{32}$, $p = e_{11}$. As before, we get $(xy)z = x(yz)$. For general x, $x + x^*$ is Hermitian, so the above equality yields

$$x(yz)+x^*(yz)=(xy)z+(x^*y)z, \qquad x, y, z\in R. \tag{2.76}$$

Let $q=e_{11}+e_{22}$, and define b and c as above. Then $b=\{qbq\}$, so using 2.4.18 we have

$$\{bcb\}=\{\{qbq\}c\{qbq\}\}=\{q\{b\{qcq\}b\}q\}=0,$$

since, in the Peirce decomposition given by e_{11}, e_{22}, e_{33}, $c\in A_{23}$, so that $\{qcq\}=0$. (This can also be seen by direct calculation.) In these calculations it must be kept in mind that, because R may not be associative, the triple product is not given by $\{bcb\}=bcb$ but rather by the definition $\{bcb\}=2b\circ(b\circ c)-b^2\circ c$. Computing the matrix products in $2b\circ(b\circ c)=b^2\circ c$, we arrive at the identity

$$x^*(xy)=(x^*x)y, \qquad x, y\in R. \tag{2.77}$$

With $y=x$, (2.76) becomes

$$x(xy)+x^*(xy)=x^2y+(x^*x)y.$$

When combined with (2.77) this yields $x(xy)=x^2y$, i.e. the left alternative law. Taking adjoints, we obtain the right alternative law.

2.7.7. When we consider $\mathbb{R}$, $\mathbb{C}$, $\mathbb{H}$ and $\mathbb{O}$ as $*$ algebras, the involution is defined to be the identity on $\mathbb{R}$, and the usual conjugation in the other cases. This involution is characterized by linearity over $\mathbb{R}$ and the requirements $1^*=1$, $i^*=-i$ whenever $i^2=-1$ (see the proof of 2.2.6).

2.7.8. Theorem. *Suppose R is a real $*$ algebra with 1 such that $R_{sa}=\mathbb{R}1$, and $x^*x\neq 0$ for all nonzero elements x of R. Suppose $H_n(R)$ is a Jordan matrix algebra, $n\geqslant 3$. Then R is $*$ isomorphic to $\mathbb{R}$, $\mathbb{C}$, $\mathbb{H}$ or $\mathbb{O}$. If $n\geqslant 4$ then R is $*$ isomorphic to $\mathbb{R}$, $\mathbb{C}$ or $\mathbb{H}$.*

Proof. We shall use 2.2.6. Hence we must prove that R is a quadratic, alternative division algebra, that R is associative if $n\geqslant 4$, and that the involution on R satisfies the requirement of 2.7.7.

By 2.7.6, R is alternative and, if $n\geqslant 4$, it is associative.

If $x\in R$ is nonzero, the assumptions imply that $x^*x=\lambda 1$, for some nonzero real number λ. Thus, $\lambda^{-1}x^*$ is a left inverse for x. Similarly x has a right inverse, so R is a division algebra.

To see that R is quadratic, let $y\in R$. Then $y=\frac{1}{2}(y+y^*)+\frac{1}{2}(y-y^*)$, or $y=\lambda 1+z$, $\lambda\in\mathbb{R}$, $z^*=-z$. Then $y^2=\lambda^2 1+z^2+2\lambda z$. Since $z^2=-z^*z\in\mathbb{R}1$, this implies that y^2 is a linear combination of 1 and z, hence of 1 and y.

Finally, if $i^2=-1$, write $i^*i=\lambda 1$ for some $\lambda\in\mathbb{R}1$. Multiplying by $-i$ from the right, we get $i^*=-\lambda i$ and taking adjoints $i=-\lambda i^*=\lambda^2 i$, so that

$\lambda = \pm 1$, or $i^* = \pm i$. If $i^* = i$ then $i \in \mathbb{R}$, which is impossible since $i^2 = -1$. Hence $i^* = -i$, so the condition of 2.7.7 is satisfied.

2.8. Coordinatization theorems

2.8.1. In this section we shall give conditions for a Jordan algebra A to be a Jordan matrix algebra, say $A \cong H_n(R)$, where R is a unital $*$ algebra and $n \geq 3$.

It is clear that a necessary condition is the existence of n pairwise orthogonal idempotents with sum 1. Further, we must have elements in the algebra corresponding to $e_{ij} + e_{ji}$, $i \neq j$. Notice that $(e_{ij} + e_{ji})^2 = e_{ii} + e_{jj}$.

Let p, q be orthogonal idempotents in a Jordan algebra A. They are said to be *strongly connected* if there is $v \in \{pAq\}$ such that $v^2 = p + q$.

Thus a necessary condition for the existence of an isomorphism $A \cong H_n(R)$ is the existence of n pairwise orthogonal and strongly connected idempotents with sum 1. One of the main objectives in this section is to prove the converse.

In proving this coordinatization theorem one easily gets bogged down in technical details. However, the whole proof is hinged on a few simple observations. Consider a Jordan matrix algebra $H_n(R)$. It contains the orthogonal idempotents $e_{11}, \ldots, e_{nn}$, with sum 1, and the elements $t_j = e_{1j} + e_{j1}$ implement the strong connectedness between e_{11} and e_{jj}, for $j = 2, \ldots, n$. From this we can recover the whole set of 'symmetrized matrix units', defined by

$$t_{ii} = e_{ii}, \qquad t_{ij} = e_{ij} + e_{ji}, \qquad i \neq j.$$

Indeed, $t_{1j} = t_j$ and $t_{ij} = \{s_i t_j s_i\}$, where $i \neq j$ and $s_i = t_i + \sum_{i \neq j \neq 1} e_{jj}$. Next, with symmetries s_{ij} defined by

$$s_{ij} = \sum_{\substack{k \neq i \\ k \neq j}} e_{kk} + t_{ij} \qquad (i \neq j),$$

we can compare elements of different Peirce components, since

$$\{s_{ij}(xe_{jk} + x^* e_{kj})s_{ij}\} = xe_{ik} + x^* e_{ki},$$

where i, j, k are distinct and $x \in R$. Thus each Peirce component A_{ij} can be identified linearly with R. Finally, the multiplication in R can be recovered by the formula

$$2(xe_{ij} + x^* e_{ji}) \circ (ye_{jk} + y^* e_{kj}) = xye_{ik} + (xy)^* e_{ki},$$

where i, j, k are distinct.

We shall first prove a coordinatization theorem for special algebras.

The result will be of independent interest, and also the proof will serve to clarify the above ideas.

2.8.2. Lemma. *Let $\mathcal{A}$ be an associative $*$ algebra with a complete set of $n \times n$ matrix units (e_{ij}), i.e. $e_{ij}e_{kl} = 0$ if $j \neq k$, $e_{ij}e_{jk} = e_{ik}$, $e_{ij}^* = e_{ji}$ and $\sum_{i=1}^n e_{ii} = 1$. Let $\mathcal{A}_0 = \{a \in \mathcal{A}: ae_{ij} = e_{ij}a$ for all i, j$\}$. Then $\mathcal{A}_0$ is a $*$ subalgebra of $\mathcal{A}$ such that the map $(a_{ij}) \to \sum_{i,j} a_{ij}e_{ij}$ is a $*$ isomorphism of $M_n(\mathcal{A}_0)$ onto $\mathcal{A}$.*

Proof. It is trivial to show that $\mathcal{A}_0$ is a $*$ subalgebra of $\mathcal{A}$. It is also easy to see that the map $\alpha: (a_{ij}) \to \sum_{i,j} a_{ij}e_{ij}$ is a $*$ homomorphism of $M_n(\mathcal{A}_0)$ into $\mathcal{A}$. To show that α is injective, assume that $\sum_{i,j} a_{ij}e_{ij} = 0$, $a_{ij} \in \mathcal{A}_0$. For any indices r, s, t we find

$$0 = e_{rs}\Big(\sum_{i,j} a_{ij}e_{ij}\Big)e_{tr} = a_{st}e_{rr}.$$

Summing over r, we get $a_{st} = 0$, proving injectivity. If $a \in \mathcal{A}$ let $a_{ij} = \sum_l e_{li}ae_{jl}$. It is easy to see that $a_{ij} \in \mathcal{A}_0$, and that $a_{ij}e_{ij} = e_{ii}ae_{jj}$, so that $\alpha((a_{ij})) = \sum a_{ij}e_{ij} = a$. In other words, α is surjective.

2.8.3. Theorem. *Let $\mathcal{A}$ be a unital associative $*$ algebra over a field Φ with involution and of characteristic not 2. Let $\Phi_0 = \{\lambda \in \Phi: \bar{\lambda} = \lambda\}$. Let $\mathcal{A}_{sa}$ be the special Jordan algebra over Φ_0 consisting of self-adjoint elements in $\mathcal{A}$, and let A be a Jordan subalgebra of $\mathcal{A}_{sa}$. Suppose A contains n mutually orthogonal and strongly connected idempotents with sum 1, where $n \geq 3$. Then there is a $*$ subalgebra $\mathcal{A}_0$ of $\mathcal{A}$, a $*$ subalgebra R of $\mathcal{A}_0$ (over Φ_0) and an isomorphism of $M_n(\mathcal{A}_0)$ onto $\mathcal{A}$ mapping $H_n(R)$ onto A.*

Proof. Let $p_1, \ldots, p_n$ be mutually orthogonal and strongly connected idempotents in A with sum 1. For $j = 2, \ldots, n$, let $t_j \in \{p_1Ap_j\}$ be such that $t_j^2 = p_1 + p_j$. Let $t_1 = p_1$, and define

$$e_{ij} = t_it_j \qquad i \neq j,\ i, j \in \{1, \ldots, n\},$$
$$e_{ii} = p_i \qquad i \in \{1, \ldots, n\}.$$

Note that the orthogonality of p_i, p_j if $i \neq j$ means $p_ip_j + p_jp_i = 0$. Multiplying from both sides with p_i we get $p_ip_jp_i = 0$, while multiplying from the left only yields $p_ip_j + p_ip_jp_i = 0$. Together these two formulae show that $p_ip_j = 0$. Symmetrically $p_jp_i = 0$.

It is now simple to show that (e_{ij}) form a complete set of $n \times n$ matrix units. Indeed, since for $i \neq 1$, $t_i = p_1t_ip_i + p_it_ip_1$, we have $e_{ij} = t_it_j = t_ip_1t_j$ when 1, i, j are distinct. Also, $p_it_i = t_ip_1$. Clearly these equalities also hold if $i \neq j$ but i or j equals 1. Hence we have $p_ie_{ij} = p_it_it_j = t_ip_1t_j = e_{ij}$, and similarly $e_{ij}p_j = e_{ij}$. Therefore, if $j \neq k$, $e_{ij}e_{kl} = e_{ij}p_jp_ke_{kl} = 0$. The formula $e_{ij}e_{jk} = e_{ik}$ follows, if $i = j$ or $j = k$, from $p_je_{jk} = e_{jk}$ or $e_{ij}p_j = e_{ij}$. Otherwise, $e_{ij}e_{jk} = t_it_j^2t_k = t_i(p_1 + p_j)t_k = t_ip_1t_k = e_{ik}$.

Evidently, $e_{ii} \in A$ and $e_{ij} + e_{ji} \in A$, if $i \neq j$.

Let $\mathscr{A}_0$ be as in 2.8.2. If $i \neq j$ let

$$R_{ij} = \{a \in \mathscr{A}_0 \colon ae_{ij} + a^* e_{ji} \in A\}.$$

We claim that all the R_{ij} are equal to the same $*$ algebra R, and that, if $a_{ij} \in \mathscr{A}_0$, $\sum_{i,j} a_{ij}e_{ij} \in A$ if and only if $a_{ij} = a_{ji}^* \in R$ for all i, j. By 2.8.2 this will complete the proof.

Let i, j, k be distinct. If $a \in R_{ij}$ then, since $ae_{ij} + a^* e_{ji} \in A$ and $e_{ii} + e_{jk} + e_{kj} \in A$,

$$ae_{ik} + a^* e_{ki} = (e_{ii} + e_{jk} + e_{kj})(ae_{ij} + a^* e_{ji})(e_{ii} + e_{jk} + e_{kj}) \tag{2.78}$$

belongs to A. Therefore, $a \in R_{ik}$. Similarly

$$a^* e_{ij} + ae_{ji} = (e_{ij} + e_{ji})(ae_{ij} + a^* e_{ji})(e_{ij} + e_{ji}) \tag{2.79}$$

belongs to A, so $a^* \in R_{ij}$, or $a \in R_{ji}$. From the properties $R_{ij} = R_{ik}$ and $R_{ij} = R_{ji}$ we conclude that all the R_{ij} $(i \neq j)$ are equal to the same Φ_0 vector space R, and also that $a \in R$ implies $a^* \in R$.

Let a, $b \in R$. Then, since $ae_{12} + a^* e_{21}$ and $be_{23} + b^* e_{32}$ belong to A, so does

$$abe_{13} + (ab)^* e_{31} = 2(ae_{12} + a^* e_{21}) \circ (be_{23} + b^* e_{32}). \tag{2.80}$$

Therefore $ab \in R$, so that R is a $*$ algebra over Φ_0.

To complete the proof of our assertion, let $x = \sum_{i,j} a_{ij}e_{ij}$ belong to A, $a_{ij} \in \mathscr{A}_0$. Since $x = x^*$, $a_{ij}^* = a_{ji}$, by 2.8.2. We must show $a_{ij} \in R$. Note first that $a_{ii}e_{ii} = e_{ii}xe_{ii} \in A$. Then if $i \neq j$

$$a_{ii}e_{ij} + a_{ii}^* e_{ji} = 2(e_{ij} + e_{ji}) \circ (a_{ii}e_{ii})$$

belongs to A, so that $a_{ii} \in R$. Next,

$$a_{ii}e_{ii} + a_{jj}e_{jj} + a_{ij}e_{ij} + a_{ji}e_{ji} = (e_{ii} + e_{jj})x(e_{ii} + e_{jj})$$

belongs to A. Since the first two terms on the left-hand side belong to A, so does $a_{ij}e_{ij} + a_{ij}^* e_{ji} = a_{ij}e_{ij} + a_{ji}e_{ji}$, so $a_{ij} \in R$.

In the converse direction, if x is given as above with $a_{ij} = a_{ji}^* \in R$, we must show that $x \in A$. For this it is sufficient to show that, for $i \neq j$, $a_{ii}e_{ii} \in A$ and $a_{ij}e_{ij} + a_{ji}e_{ji} \in A$. The latter statement is true by definition of R. Further, we have $a_{ii} = a_{ii}^* \in R$, so $a_{ii}e_{ij} + a_{ii}e_{ji} \in A$. This implies that

$$a_{ii}(e_{ii} + e_{jj}) = (a_{ii}e_{ij} + a_{ii}e_{ji}) \circ (e_{ij} + e_{ji})$$

is in A, and, therefore, so is $a_{ii}e_{ii} = (a_{ii}(e_{ii} + e_{jj})) \circ e_{ii}$. This completes the proof of our assertion, and hence the theorem.

2.8.4. Corollary. *If A is a special, unital Jordan algebra containing $n \geq 3$ pairwise orthogonal and strongly connected idempotents with sum 1, then $A \cong H_n(R)$ for some associative $*$ algebra R.*

Proof. Let $\phi: A \to B^J$ be an injective homomorphism, where B is an associative algebra. Let B^0 be the algebra opposite to B; it is characterized by the existence of an anti-isomorphism $a \to a^0$ of B onto B^0. $B \oplus B^0$ is a $*$ algebra with involution $(a \oplus b^0)^* = b \oplus a^0$. Then ψ: $a \to \phi(a) \oplus \phi(a)^0$ is an injective homomorphism of A into the Hermitian part of $B \oplus B^0$. Let $\mathcal{A}$ be the $*$ subalgebra $\psi(1)(B \oplus B^0)\psi(1)$ of $B \oplus B^0$. Again ψ: $A \to \mathcal{A}_{sa}$ is injective, and $\psi(1)$ is the unit of $\mathcal{A}$. Theorem 2.8.3 can now be applied to $\mathcal{A}$ and its Jordan subalgebra $\psi(A)$.

2.8.5. Corollary. *$H_3(\mathbb{O})$ is exceptional.*

Proof. Assume $H_3(\mathbb{O})$ is special. The idempotents e_{11}, e_{22}, e_{33} in $H_3(\mathbb{O})$ are orthogonal with sum 1. They are also strongly connected, via $e_{ij} + e_{ji}$, $i \neq j$. Working through the proof of 2.8.3 with $t_j = e_{1j} + e_{j1}$, $j = 2, 3$, we get a real, associative $*$ algebra R and an isomorphism α: $H_3(\mathbb{O}) \to H_3(R)$ mapping e_{ii} to e_{ii} and $e_{ij} + e_{ji}$ to $e_{ij} + e_{ji}$, when $i \neq j$.

If $i \neq j$ and $x \in \mathbb{O}$, consider $a = xe_{ij} + x^* e_{ji} \in \{e_{ii} H_3(\mathbb{O}) e_{jj}\}$. Since then $\alpha(a) \in \{e_{ii} H_3(R) e_{jj}\}$, we have

$$\alpha(xe_{ij} + x^* e_{ji}) = \phi_{ij}(x) e_{ij} + \phi_{ij}(x)^* e_{ji},$$

where $\phi_{ij}: \mathbb{O} \to R$ is a real linear isomorphism. By (2.78) and (2.79) ϕ_{ij} is independent of i, j, so we drop the indices and call it ϕ. Using (2.80) we see that $\phi(xy) = \phi(x)\phi(y)$, so $\mathbb{O}$ and R are isomorphic as algebras. However, since $\mathbb{O}$ is not associative (see e.g. (2.16)), this is absurd. This contradiction completes the proof.

2.8.6. Let A be a unital Jordan algebra, and $s \in A$ a symmetry, i.e. $s^2 = 1$. Then U_s is an automorphism of A, for by Macdonald's theorem we have $\{cac\} \circ \{cbc\} = \{c\{ac^2b\}c\}$ for any a, b, c in a Jordan algebra, and with $c = s$ this is $(U_s a) \circ (U_s b) = U_s(a \circ b)$.

2.8.7. In a Jordan matrix algebra $H_n(R)$ consider the symmetry

$$s_{ij} = e_{ij} + e_{ji} + \sum_{k \neq i,j} e_{kk}$$

for $i \neq j$. The corresponding automorphism, $U_{(ij)}$: $a \to \{s_{ij} a s_{ij}\}$, acts on a matrix simply by interchanging the indices i and j. Thus, all the automorphisms $U_{(ij)}$ generate a group isomorphic to the full symmetric group S_n on n elements. This will be our clue to the general coordinatization theorem. The following theorem on generators of S_n will be needed.

2.8.8. Theorem. *Let $F(x_2, \ldots, x_n)$ be the free group on $n-1$ generators, and let*

$$G_n = F(x_2, \ldots, x_n)/(x_i^2, (x_i x_j)^3, (x_i x_j x_i x_k)^2; i, j, k \text{ distinct}),$$

in other words G_n is the quotient by the normal subgroup generated by the indicated elements. Then G_n is isomorphic to the symmetric group S_n on $\{1, \ldots, n\}$ via an isomorphism mapping x_i to the transposition $(1i)$, $i = 2, \ldots, n$.

Proof. Let us rewrite the relationships defining G_n. $(x_i x_j)^3 = 1$ is equivalent to $(x_i x_j x_i)(x_j x_i x_j) = 1$. Since $x_i^2 = 1$, $x_i x_j x_i$ is its own inverse, so the above is equivalent to

$$x_i x_j x_i = x_j x_i x_j. \tag{2.81}$$

Similarly, the inverse of $x_i x_j x_i x_k$ is $x_k x_i x_j x_i$, so $(x_i x_j x_i x_k)^2 = 1$ is equivalent to

$$(x_i x_j x_i) x_k = x_k (x_i x_j x_i). \tag{2.82}$$

Thus G_n is the group defined by the relations $x_i^2 = 1$, (2.81) and (2.82).

For permutations we have $(1i)^2 = 1$, and if i, j, $k \in \{2, \ldots, n\}$ are distinct,

$$(1i)(1j)(1i) = (ij) = (1j)(1i)(1j),$$
$$[(1i)(1j)(1i)](1k) = (ij)(1k) = (1k)(ij) = (1k)[(1i)(1j)(1i)],$$

so $(1i)$, $i = 2, \ldots, n$, satisfy the relations defining G_n. Therefore, there is a homomorphism π: $G_n \to S_n$ such that $\pi(x_i) = (1i)$. Clearly, π maps G_n onto S_n.

Let H be the subgroup of G_n generated by $x_2, \ldots, x_{n-1}$. We assert that the index satisfies

$$(G_n : H) \leq n. \tag{2.83}$$

Suppose for the moment that (2.83) is true. Notice that H is a quotient of G_{n-1} if $n \geq 3$. Therefore,

$$(G_n : 1) = (G_n : H)(H : 1) \leq n(G_{n-1} : 1).$$

Since G_2 has two elements, $(G_n : 1) \leq n!$ follows by induction. However, since S_n is a quotient of G_n and S_n has $n!$ elements, it follows that π is an isomorphism, which was to be proved.

To motivate the proof of (2.83), we point out that S_n/S_{n-1} can be identified with $\{1, \ldots, n\}$ via the map $j \to (jn)S_{n-1}$, $j = 1, \ldots, n-1$, $n \to S_{n-1}$. We ought to have $H \cong S_{n-1}$, so by analogy we should be able to guess at representatives for the n cosets in G_n/H. Thus, let

$$s_1 = x_n H, \qquad s_2 = x_2 x_n H, \qquad \ldots, \qquad s_{n-1} = x_{n-1} x_n H, \qquad s_n = H.$$

We shall show that, under the action of G_n on G_n/H, each x_i maps $\{s_1, \ldots, s_n\}$ into itself. Then, since G_n acts transitively on G_n/H, (2.83) follows.

First consider the action of x_n. Clearly, $x_n s_1 = s_n$, $x_n s_n = s_1$. For $2 \leqslant i \leqslant n-1$ we get, using (2.81), $x_n s_i = x_n x_i x_n H = x_i x_n x_i H = x_i x_n H = s_i$. Next, let $2 \leqslant i \leqslant n-1$ and consider the action of x_i. Clearly $x_i s_i = s_1$, $x_i s_1 = s_i$, $x_i s_n = s_n$. For $2 \leqslant j \leqslant n-1$ and $j \neq i$ we find, using (2.82), $x_i s_j = x_i x_j x_n H = x_i x_j x_n x_j H = x_j x_n x_j x_i H = x_j x_n H = s_j$. This completes the proof of (2.83) and hence of the theorem.

2.8.9. The coordinatization theorem. *Let A be a unital Jordan algebra. Suppose that A contains $n \geqslant 3$ pairwise orthogonal strongly connected idempotents with sum 1. Then A is isomorphic to $H_n(R)$ for some * algebra R.*

2.8.10. The proof of 2.8.9 will be broken into a series of lemmas. We first fix some notation to be kept throughout the proof. Let $p_1, \ldots, p_n$ be the idempotents mentioned in 2.8.9. Let A_{ij} be the Peirce component $A_{ij} = \{p_i A p_j\}$. For each j, $2 \leqslant j \leqslant n$, let $t_j \in A_{1j}$ satisfy $t_j^2 = p_1 + p_j$. (All we shall need to prove 2.8.9 is the existence of the t_j, i.e. that p_1 and p_j are strongly connected.) Further, we fix the following notation:

$$t_{ii} = p_i \qquad (i = 1, \ldots, n),$$

$$t_{ij} = 2t_i \circ t_j \qquad (i \neq j),$$

$$s_{ij} = t_{ij} + \sum_{k \neq i,j} t_{kk}.$$

For the definition of t_{ij} to work in all cases we let $t_1 = p_1$, so that $t_{1j} = t_{j1} = t_j$.

We shall need the following two identities, of which the second follows from the first, and both follow from Macdonald's theorem:

$$4(a \circ b) \circ (a \circ c) = \{aba\} \circ c + \{aca\} \circ b + \{a(b \circ c)a\} + \{ba^2c\}, \qquad (2.84)$$

$$4(a \circ b)^2 = 2\{aba\} \circ b + \{ab^2a\} + \{ba^2b\}. \qquad (2.85)$$

2.8.11. Lemma. *The t_{ij} form a set of symmetrized matrix units,* i.e. *they multiply like $t'_{ii} = e_{ii}$, $t'_{ij} = e_{ij} + e_{ji}$:*

(i) $t_{ii}^2 = t_{ii}$, $\sum_{i=1}^n t_{ii} = 1$;
(ii) $t_{ii} \circ t_{ij} = \frac{1}{2} t_{ij}$ $(i \neq j)$;
(iii) $t_{ij}^2 = t_{ii} + t_{jj}$ $(i \neq j)$;
(iv) $t_{ij} \circ t_{jk} = \frac{1}{2} t_{ik}$ $(i \neq k)$;
(v) $t_{ij} \circ t_{kl} = 0$ $(\{i, j\} \cap \{k, l\} = \varnothing)$.

Proof. (i) is trivial.

(ii) is also trivial, since $t_i \in A_{1i}$, $t_j \in A_{1j}$ implies $t_{ij} = 2t_i \circ t_j \in A_{ij}$ (see 2.6.5).

(iii) If $i = 1$ or $j = 1$ (iii) follows immediately from the assumption

$t_i^2 = p_1 + p_i$. So we may assume that 1, i, j are all distinct. By 2.6.6 $\{A_{1i}A_{1j}A_{1i}\} = 0$, so by (2.85) we get

$$\begin{aligned} t_{ij}^2 = 4(t_i \circ t_j)^2 &= 2\{t_it_jt_i\} \circ t_j + \{t_it_j^2t_i\} + \{t_jt_i^2t_j\} \\ &= 0 + \{t_i(p_1+p_j)t_i\} + \{t_j(p_1+p_i)t_j\} \\ &= \{t_ip_1t_i\} + \{t_jp_1t_j\}, \end{aligned}$$

since $\{t_ip_jt_i\} = \{t_jp_it_j\} = 0$ by 2.6.5 (and the definition of the triple product). Furthermore, we get $\{t_ip_1t_i\} = 2t_i \circ (t_i \circ p_1) - t_i^2 \circ p_1 = t_i^2 - t_i^2 \circ p_1 = (p_1 + p_i) \circ (1 - p_1) = p_i = t_{ii}$ and similarly $\{t_jp_1t_j\} = t_{jj}$. Thus (iii) follows.

(iv) If $i = j$ or $j = k$ this is (ii), so we may assume that i, j, k are all distinct. If $j = 1$ then $t_{ij} = t_i$, $t_{jk} = t_k$, so (iv) is nothing but the definition of t_{ik}. If $i = 1$ we find, since as above $\{t_jt_kt_j\} = 0$, that

$$\begin{aligned} t_{1j} \circ t_{jk} &= 2t_j \circ (t_j \circ t_k) = \{t_jt_kt_j\} + t_j^2 \circ t_k \\ &= 0 + (p_1 + p_j) \circ t_k = \tfrac{1}{2}t_{1k}. \end{aligned}$$

If $k = 1$ we prove (iv) similarly. We may therefore assume that 1, i, j, k are all distinct. Using (2.84) we get

$$\begin{aligned} t_{ij} \circ t_{jk} &= 4(t_i \circ t_j) \circ (t_j \circ t_k) \\ &= \{t_jt_it_j\} \circ t_k + \{t_jt_kt_j\} \circ t_i + \{t_j(t_i \circ t_k)t_j\} + \{t_it_j^2t_k\}. \end{aligned}$$

The first two terms vanish by 2.6.6 and the third by 2.6.5, since $t_i \circ t_k \in A_{ik}$. Hence

$$\begin{aligned} t_{ij} \circ t_{jk} &= \{t_i(p_1 + p_j)t_k\} = \{t_ip_1t_k\} \\ &= (t_i \circ p_1) \circ t_k + (t_k \circ p_1) \circ t_i - (t_i \circ t_k) \circ p_1 \\ &= \tfrac{1}{2}t_i \circ t_k + \tfrac{1}{2}t_k \circ t_i - 0 \\ &= t_i \circ t_k = \tfrac{1}{2}t_{ik}. \end{aligned}$$

Finally, (v) is obvious from 2.6.5.

2.8.12. From 2.8.11 it is immediately clear that $s_{ij} = t_{ij} + \sum_{k \neq i,j} t_{kk}$ is a symmetry. From the remarks in 2.8.7 we expect the automorphism $U_{s_{ij}}$ to have no effect other than interchanging i, j in our matrix algebra representation of A.

2.8.13. Lemma. *Let $i \neq j$. Then $U_{s_{ij}}$ is the identity on A_{kl} if $\{i, j\} \cap \{k, l\} = \varnothing$. It interchanges A_{ii} and A_{jj} and also A_{ik} and A_{jk} if $k \neq i, j$. It maps A_{ij} into itself. Moreover if $k \neq i, j$ then*

$$\begin{aligned} &U_{s_{ij}}(t_{ii}) = t_{jj}; \\ &U_{s_{ij}}(t_{ij}) = t_{ij}; \qquad U_{s_{ij}}(s_{ij}) = s_{ij}; \\ &U_{s_{ij}}(t_{jk}) = t_{ik}; \qquad U_{s_{ij}}(s_{jk}) = s_{ik}. \end{aligned}$$

Proof. If we remember that $t_{ij} \in A_{ij}$ and the rules for multiplication of Peirce components, the first statement is easy. Next we find

$$\begin{aligned} U_{s_{ij}}(t_{ii}) &= \{t_{ij}t_{ii}t_{ij}\} = 2t_{ij}\circ(t_{ij}\circ t_{ii}) - t_{ij}^2 \circ t_{ii} \\ &= t_{ij}\circ t_{ij} - (t_{ii}+t_{jj})\circ t_{ii} \\ &= t_{ii} + t_{jj} - t_{ii} = t_{jj}. \end{aligned}$$

Symmetrically $U_{s_{ij}}(t_{jj}) = t_{ii}$. By the first part of the proof $U_{s_{ij}}(t_{kk}) = t_{kk}$ if $k \neq i, j$. From this the statements on Peirce components follows, since $U_{s_{ij}}$ is an automorphism. Indeed, any automorphism α maps $\{pAq\}$ onto $\{\alpha(p)A\alpha(q)\}$.

It is immediate that

$$U_{s_{ij}}(t_{ij}) = t_{ij}^3 = (t_{ii}+t_{jj})\circ t_{ij} = \tfrac{1}{2}t_{ij} + \tfrac{1}{2}t_{ij} = t_{ij},$$

and, even more trivially, $U_{s_{ij}}(s_{ij}) = s_{ij}$. Next if $k \neq i, j$ we have

$$\begin{aligned} U_{s_{ij}}(t_{jk}) &= \left\{\left(t_{ij} + \sum_{l\neq i,j} t_{ll}\right) t_{jk} \left(t_{ij} + \sum_{m\neq i,j} t_{mm}\right)\right\} \\ &= \{t_{ij}t_{jk}t_{ij}\} + 2\sum_{l\neq i,j}\{t_{ll}t_{jk}t_{ij}\} + \sum_{l\neq i,j}\sum_{m\neq i,j}\{t_{ll}t_{ij}t_{mm}\}. \end{aligned}$$

Here the first term vanishes by 2.6.6(ii). By the multiplication rules for Peirce components each term of the double sum vanishes, and so do all terms of the middle sum except the $l = k$ term. Hence by 2.8.11(iv),

$$\begin{aligned} U_{s_{ij}}(t_{jk}) &= 2\{t_{kk}t_{jk}t_{ij}\} \\ &= 2[(t_{ij}\circ t_{jk})\circ t_{kk} + (t_{kk}\circ t_{jk})\circ t_{ij} - (t_{ij}\circ t_{kk})\circ t_{jk}] \\ &= t_{ik}\circ t_{kk} + t_{jk}\circ t_{ij} - 0 \\ &= \tfrac{1}{2}t_{ik} + \tfrac{1}{2}t_{ik} = t_{ik}. \end{aligned}$$

The equation $U_{s_{ij}}(s_{jk}) = s_{ik}$ follows, since $U_{s_{ij}}$ is an automorphism and $s_{jk} = t_{jk} + 1 - t_{jk}^2$.

2.8.14. Lemma. *There is a group isomorphism $\pi \to U_\pi$ of the symmetric group S_n on $\{1, \ldots, n\}$ into the automorphism group* Aut(A) *such that $U_{(ij)} = U_{s_{ij}}$ if $i \neq j$. Moreover, if $\pi \in S_n$ then $U_\pi(A_{ij}) = A_{\pi(i)\pi(j)}$, $U_\pi(t_{ij}) = t_{\pi(i)\pi(j)}$. $U_\pi(s_{ij}) = s_{\pi(i)\pi(j)}$ for all i, j. Finally, the restriction of U_π to A_{ij} depends only on $\pi(i)$ and $\pi(j)$.*

Proof. We must first show that the operators $U_{s_{1i}}$, $i = 2, \ldots, n$, satisfy the relations of 2.8.8. First, for any symmetry s we have $U_s^2 = U_{s^2} = 1$. Next, by 2.8.13 the identity $\{a\{b\{aca\}b\}a\} = \{\{aba\}c\{aba\}\}$ (2.4.18) implies

$$U_{s_{1i}}U_{s_{1j}}U_{s_{1i}} = U_{\{s_{1i}s_{1j}s_{1i}\}} = U_{s_{ij}} \tag{2.86}$$

if $i \neq j$. In particular, by symmetry

$$U_{s_{1i}} U_{s_{1j}} U_{s_{1i}} = U_{s_{1j}} U_{s_{1i}} U_{s_{1j}},$$

which is the relation (2.81). Also, if i, j, k are distinct in $\{2, \ldots, n\}$ then, using 2.6.3 with $p = t_{ii} + t_{jj}$, we see that t_{ij} and t_{1k} operator commute. Since they both operator commute with $t_{ii} + t_{jj}$ and $t_{11} + t_{kk}$ and with $\sum_{l \neq 1,i,j,k} t_{ll}$, we see that s_{ij} and s_{1k} operator commute. Hence $U_{s_{ij}} U_{s_{1k}} = U_{s_{1k}} U_{s_{ij}}$, and using (2.86) we find that the U_{1i} satisfy relation (2.82).

From 2.8.8 it now follows that there is a homomorphism $\pi \to U_\pi$ of S_n into Aut(A) such that $U_{(1i)} = U_{s_{1i}}$, $i = 1, \ldots, n$. From (2.86) and the formula $(ij) = (1i)(1j)(1i)$ we see that $U_{(ij)} = U_{s_{ij}}$.

The formulae $U_\pi(A_{ij}) = A_{\pi(i)\pi(j)}$, $U_\pi(t_{ij}) = t_{\pi(i)\pi(j)}$ and $U_\pi(s_{ij}) = s_{\pi(i)\pi(j)}$ are proved in 2.8.13 for a generating set $(12), \ldots, (1n)$, so it follows for all π. Evidently, if $U_\pi = 1$ it follows from this that $\pi = 1$, so $\pi \to U_\pi$ is an isomorphism into Aut(A).

To prove the final statement it is enough to show that if $\pi(i) = i$, $\pi(j) = j$ then the restriction of U_π to A_{ij} is the identity. But π is then a product of transpositions (kl) with $\{k, l\} \cap \{i, j\} = \varnothing$, so this also follows from 2.8.13.

2.8.15. We are now in a position to define the $*$ algebra R which will satisfy $A = H_n(R)$.

Let R be the set of all indexed families $x = (x_{ij})_{i \neq j}$, where $x_{ij} \in A_{ij}$, subject to the condition

$$U_\pi(x_{ij}) = x_{\pi(i)\pi(j)} \tag{2.87}$$

for all $\pi \in S_n$, $i \neq j$.

We remark here that $x \to x_{ij}$ is a linear isomorphism of R onto A_{ij} for $i \neq j$; hence all the 'off-diagonal' Peirce components A_{ij} are isomorphic to each other and to R. However, these isomorphisms are not canonical, because x_{ji} may be different from x_{ij}. This gives rise to our definition of the adjoint: if $x \in R$ define

$$x^*_{ij} = x_{ji} \qquad (i \neq j). \tag{2.88}$$

It is clear that $x^* \in R$.

Similarly, we define a product in R as follows: if $x, y \in R$ let

$$(xy)_{ij} = 2x_{ik} \circ y_{kj} \qquad (i, j, k \text{ distinct}). \tag{2.89}$$

We must show three things: $(xy)_{ij}$ as defined above does not depend on k, $U_\pi((xy)_{ij}) = (xy)_{\pi(i)\pi(j)}$ and $(xy)_{ij} \in A_{ij}$. The latter follows from the Peirce multiplication rules (2.6.5), and the former from the following computation, in which we use that U_π is an automorphism:

$$\begin{aligned} U_\pi(x_{ik} \circ y_{kj}) &= U_\pi(x_{ik}) \circ U_\pi(y_{kj}) \\ &= x_{\pi(i)\pi(k)} \circ y_{\pi(k)\pi(j)}. \end{aligned}$$

Finally, if we choose π such that $\pi(i)=i$, $\pi(j)=j$ we see that the definition (2.89) does not depend on k, and then a general choice of π shows $xy \in R$.

2.8.16. Lemma. *R is a $*$ algebra with unit $1=(t_{ij})_{i\neq j}$.*

Proof. From (2.88) it is clear that $x^{**}=x$. From (2.88) and (2.89) we get

$$\begin{aligned}(xy)^*_{ij} &= (xy)_{ji} = 2x_{jk} \circ y_{ki} \\ &= 2x^*_{kj} \circ y^*_{ik} = 2y^*_{ik} \circ x^*_{kj} = (y^*x^*)_{ij},\end{aligned}$$

so $(xy)^* = y^*x^*$, i.e. R is a $*$ algebra.

Define $t=(t_{ij})_{i\neq j}$. Clearly, by (2.87) and 2.8.14, $t \in R$. We must show that t is a (right) identity for R. Since $t=t^*$, it will also be a left identity.

By the Shirshov–Cohn theorem (2.4.14) any symmetry s satisfies $U_s(a)\circ s = a \circ s$. From this we get, with i, j, k distinct,

$$x_{ij} \circ s_{jk} = U_{(jk)}(x_{ij}) \circ s_{jk} = x_{ik} \circ s_{jk}.$$

Also,

$$x_{ij} \circ s_{jk} = x_{ij} \circ t_{jk} + \sum_{l\neq j,k} x_{ij} \circ t_{ll} = x_{ij} \circ t_{jk} + \tfrac{1}{2}x_{ij},$$

$$x_{ik} \circ s_{jk} = x_{ik} \circ t_{jk} + \tfrac{1}{2}x_{ik}.$$

Comparing the above three equalities we obtain

$$x_{ik} \circ t_{jk} + \tfrac{1}{2}x_{ik} = x_{ij} \circ t_{jk} + \tfrac{1}{2}x_{ij}.$$

The terms of the above equality belong to the Peirce components A_{ij}, A_{ik}, A_{ik} and A_{ij} respectively, in that order. Hence we must have $x_{ik} \circ t_{jk} = \frac{1}{2}x_{ij}$, or $xt = x$. This completes the proof.

From now on, the identity of R will be denoted by 1.

2.8.17. If it had not been for the 'diagonal' $\sum A_{ii}$ of A, our proof of the coordinatization theorem would now almost be finished. We must show that each A_{ii} is canonically isomorphic to R_{sa} and that we get formulae like (2.89) when for example $x=x^*$ and $i=k$.

2.8.18. Lemma. *If $i\neq j$ the map $A_{ii} \to A_{ij}$ given by $a \to 2a\circ t_{ij}$ induces a linear isomorphism of A_{ii} onto R_{sa} which is independent of j. The inverse map, denoted $x \to x_{ii}$, satisfies $U_\pi(x_{ii}) = x_{\pi(i)\pi(i)}$, and $x_{ij} \circ t_{ij} = x_{ii} + x_{jj}$.*

Proof. If $a \in A_{ii}$, then $a \circ t_{ij} \in A_{ij}$, so there is some $x \in R$ such that $x_{ij} = 2a \circ t_{ij}$. It is the map $a \to x$ that is claimed to be an isomorphism of A_{ii} onto R_{sa}. From the identity $U_s(a\circ s) = a \circ s$, valid for any symmetry s, we get for any $a \in A_{ii}$ that

$$U_{(ij)}(a \circ t_{ij}) = U_{(ij)}(a \circ s_{ij}) = a \circ s_{ij} = a \circ t_{ij}. \tag{2.90}$$

Therefore, if $x \in R$ is such that $x_{ij} = 2a \circ t_{ij}$, then $x_{ji} = U_{(ij)} x_{ij} = x_{ij}$, so $x = x^*$. In other words, we have a linear map $\alpha_i \colon A_{ii} \to R_{sa}$ so that $\alpha_i(a)_{ij} = 2a \circ t_{ij}$.

To see that this definition is independent of j, note that

$$U_\pi(2a \circ t_{ij}) = 2U_\pi(a) \circ t_{\pi(i)\pi(j)} \tag{2.91}$$

for any permutation π. Choosing π such that $\pi(i) = i$, $\pi(j) = k$, we conclude that the above definition of α_i is independent of j. By choosing a more general permutation π in (2.91), we see from (2.87) that the diagram

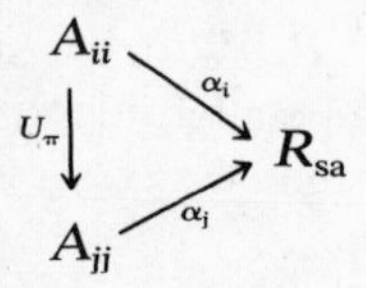

commutes, where $\pi(i) = j$.

Our next task will be to construct the map $\beta_i \colon R_{sa} \to A_{ii}$ which will be inverse to α_i. If $i \neq j$ and $x \in R_{sa}$ then, since $x_{ij} \in A_{ij}$, we get from the multiplication rules for Peirce components that

$$x_{ij} \circ t_{ij} = x_{ii} + x_{jj}, \tag{2.92}$$

where $x_{ii} \in A_{ii}$, $x_{jj} \in A_{jj}$. *A priori*, this definition of x_{ii} depends on j, but if π is a permutation such that $\pi(i) = i$, $\pi(j) = k$, we can apply U_π to (2.92) and find

$$x_{ik} \circ t_{ik} = x_{ii} + U_\pi(x_{jj}), \tag{2.93}$$

since U_π is the identity on A_{ii} (2.8.14). Furthermore, $U_\pi(x_{jj}) \in A_{kk}$, so the definition of x_{ii} is the same when k is substituted for j in (2.92). It should be noted that, since $x^* = x$, $x_{ij} = x_{ji}$. Therefore the definition (2.92) is symmetric in i and j. We write $\beta_i(x) = x_{ii}$.

From (2.93) we get $U_\pi(x_{jj}) = x_{\pi(j)\pi(j)}$, which is part of the conclusion. Formula (2.92) is another part of the conclusion, so we are left with showing that α_i and β_i are inverses of each other.

First, let $x \in R_{sa}$. Then, from the definition of the Jordan triple product together with the multiplication rules for Peirce components, we obtain

$$\begin{aligned}(x_{ij} \circ t_{ij}) \circ t_{ij} &= \tfrac{1}{2}(U_{t_{ij}} + T_{t_{ij}^2}) x_{ij} \\ &= \tfrac{1}{2}(U_{(ij)} + T_{t_{ii}+t_{jj}}) x_{ij} \\ &= \tfrac{1}{2}(x_{ji} + x_{ij}) = x_{ij}.\end{aligned}$$

Substituting (2.92) in this we have

$$(x_{ii} + x_{jj}) \circ t_{ij} = x_{ij}.$$

However, using (2.90) we have

$$x_{ii} \circ t_{ij} = U_{(ij)}(x_{ii} \circ t_{ij}) = x_{jj} \circ t_{ij}.$$

From the last two formulae we thus have $2x_{ii} \circ t_{ij} = x_{ij}$, so that $\alpha_i(\beta_i(x)) = x$.

Finally, let $a \in A_{ii}$. Again, from the definition of the triple product we have

$$\begin{aligned} 2(a \circ t_{ij}) \circ t_{ij} &= (U_{t_{ij}} + T_{t_{ij}^2})a \\ &= U_{(ij)}a + a, \end{aligned}$$

and, noting that $a \in A_{ii}$, $U_{(ij)}a \in A_{jj}$, we get that $\beta_i(\alpha_i(a)) = a$.

2.8.19. Proof of the coordinatization theorem (2.8.9). We shall define $\beta: H_n(R) \to A$ by

$$\beta\Big(\sum_{i,j} a_{ij}e_{ij}\Big) = \sum_{i \leq j} (a_{ij})_{ij} \qquad (a_{ij} = a_{ji}^* \in R). \tag{2.94}$$

This notation may look confusing; however, note that each matrix entry $a_{ij} \in R$ is itself an indexed family $((a_{ij})_{kl})_{k \neq l}$, where $(a_{ij})_{kl} \in A_{kl}$. $(a_{ij})_{ij}$ is simply the (i, j) entry of this family. If $i = j$, $(a_{ij})_{ij}$ is defined by 2.8.18.

Note that β maps the Peirce component $\{xe_{ij} + x^*e_{ji} : x \in R\}$ of $H_n(R)$ linearly isomorphically onto A_{ij}, whenever $i \neq j$. (It was noted in 2.8.15 that $x \to x_{ij}$ is a linear isomorphism of R onto A_{ij}.) Also, β maps the Peirce component $\{xe_{ii} : x \in R_{sa}\}$ linearly isomorphically onto A_{ii}, by 2.8.18. Hence β is a linear isomorphism of $H_n(R)$ onto A.

It remains to show that β preserves products. By bilinearity it is sufficient to show this when each of the factors is of the form xe_{ii} $(x = x^*)$, or $xe_{ij} + x^*e_{ji}$ $(i \neq j)$. This is clear if the two factors do not involve the same index; the product will then be zero in both A and $H_n(R)$. In other words, we must show that $\beta(a \circ b) = \beta(a) \circ \beta(b)$ in the following cases:

(i) $a = xe_{ii}$, $b = ye_{ii}$, $x, y \in R_{sa}$,
(ii) $a = xe_{ii}$, $b = ye_{ij} + y^*e_{ji}$, $x \in R_{sa}$, $y \in R$, $i \neq j$,
(iii) $a = xe_{ij} + x^*e_{ji}$, $b = ye_{jk} + y^*e_{kj}$, $x, y \in R$, i, j, k distinct,
(iv) $a = xe_{ij} + x^*e_{ji}$, $b = ye_{ij} + y^*e_{ji}$, $x, y \in R$, $i \neq j$.

Case (iii) immediately reduces to the formula $(xy)_{ik} = 2x_{ij} \circ y_{jk}$, which holds by definition (2.89). In cases (i) and (iv) we may by polarization assume $x = y$ (i.e. $a \circ b = \frac{1}{2}[(a+b)^2 - a^2 - b^2]$). Therefore the cases (i), (ii) and (iv) reduce to the following formulae:

$$(x_{ii})^2 = (x^2)_{ii} \qquad (x = x^*), \tag{2.95}$$

$$x_{ii} \circ y_{ij} = \tfrac{1}{2}(xy)_{ij} \qquad (x = x^*, i \neq j), \tag{2.96}$$

$$(x_{ij})^2 = (xx^*)_{ii} + (x^*x)_{jj} \qquad (i \neq j). \tag{2.97}$$

We prove (2.97) first. Let i, j, k be distinct. In the following computation we start out using 2.8.18, and then use the definition of multiplication in R, the fact that $1=(t_{ij})_{i\neq j}$ is the unit of R, and the definition of the triple product:

$$\begin{aligned}(xx^*)_{ii}+(xx^*)_{kk} &= t_{ik}\circ(xx^*)_{ik} = 2t_{ik}\circ(x_{ij}\circ x_{kj})\\ &= 2(-\{x_{ij}t_{ik}x_{kj}\}+(x_{ij}\circ t_{ik})\circ x_{kj}+(x_{kj}\circ t_{ik})\circ x_{ij})\\ &= -2\{x_{ij}t_{ik}x_{kj}\}+x_{kj}^2+x_{ij}^2.\end{aligned}$$

Consider the A_{ii} component of the terms in this equality. By 2.6.6(i), $\{x_{ij}t_{ik}x_{kj}\}\in A_{jj}$. Also, $x_{kj}^2\in A_{kk}+A_{jj}$, and $x_{ij}^2\in A_{ii}+A_{jj}$. It follows that

$$x_{ij}^2=(xx^*)_{ii}+b,$$

for some $b\in A_{jj}$. Applying this to x^* with i and j interchanged, we get $x_{ij}^2=(x^*)_{ji}^2=(x^*x)_{jj}+c$, for some $c\in A_{ii}$. Comparing this with the above we obtain (2.97).

We next prove (2.95). If $x=x^*$ we get from (2.97) that

$$x_{ij}^2=(x^2)_{ii}+(x^2)_{jj}.$$

In the following computation we use this, 2.8.18 and (2.85). Let $i\neq j$:

$$\begin{aligned}(x_{ii})^2+(x_{jj})^2 &= (x_{ii}+x_{jj})^2=(t_{ij}\circ x_{ij})^2\\ &= \tfrac{1}{4}[2\{t_{ij}x_{ij}t_{ij}\}\circ x_{ij}+\{t_{ij}x_{ij}^2t_{ij}\}+\{x_{ij}t_{ij}^2x_{ij}\}]\\ &= \tfrac{1}{4}[2x_{ij}^2+\{t_{ij}((x^2)_{ii}+(x^2)_{jj})t_{ij}\}+x_{ij}^2]\\ &= \tfrac{1}{4}[3x_{ij}^2+(x^2)_{jj}+(x^2)_{ii}]\\ &= (x^2)_{ii}+(x^2)_{jj}.\end{aligned}$$

If we look at the A_{ii} component of this, we have (2.95).

Finally, to prove (2.96) let $x=x^*$, $y\in R$, and let i, j and k be distinct. From 2.8.18 and the definition of the triple product,

$$\begin{aligned}x_{ii}\circ y_{ij} &= (x_{ii}+x_{kk})\circ y_{ij}=(x_{ik}\circ t_{ik})\circ y_{ij}\\ &= -\{t_{ik}y_{ij}x_{ik}\}+(x_{ik}\circ y_{ij})\circ t_{ik}+(y_{ij}\circ t_{ik})\circ x_{ik}.\end{aligned}$$

By 2.6.6(ii) the triple product vanishes. Therefore,

$$\begin{aligned}x_{ii}\circ y_{ij} &= \tfrac{1}{2}(xy)_{kj}\circ t_{ik}+\tfrac{1}{2}y_{kj}\circ x_{ik}\\ &= \tfrac{1}{4}(xy)_{ij}+\tfrac{1}{4}(xy)_{ij}=\tfrac{1}{2}(xy)_{ij},\end{aligned}$$

which is (2.96). The proof of the coordinatization theorem is complete.

2.8.20. For later reference we note that we have proved the following. If p_1, p_2 and p_3 are pairwise orthogonal idempotents with p_1 strongly connected to both p_2 and p_3, then p_2 and p_3 are strongly connected. For, assuming $p_1+p_2+p_3=1$, as we may, the proof of 2.8.11 is still valid, and in the notation of that lemma, $t_{23}\in\{p_2Ap_3\}$ and $t_{23}^2=p_2+p_3$, i.e. p_2 and p_3 are strongly connected.

2.9. Formally real Jordan algebras

2.9.1. We shall now consider Jordan algebras over the field $\mathbb{R}$ of real numbers. Such an algebra A is called *formally real* if $a_i\in A$, $\sum_{i=1}^n a_i^2=0$ implies $a_1=\ldots=a_n=0$. To justify our interest in this notion note that the Jordan algebra of all self-adjoint operators on a Hilbert space is formally real. In fact, the JB algebras to be introduced below will be formally real (3.3.8) and it will be shown later that for finite-dimensional algebras the converse is true (3.1.7).

This section is concerned with the classification of finite-dimensional, formally real Jordan algebras. These will turn out to be mostly direct sums of matrix algebras $H_n(R)$, with $R=\mathbb{R}$, $\mathbb{C}$, $\mathbb{H}$ or $\mathbb{O}$. In order to prove this we need minimal idempotents. These are defined in any ring or algebra to be nonzero idempotents p with the property that, for any nonzero idempotent q, $pq=qp=q$ implies $q=p$. Proposition 2.9.3, which incidentally characterizes those Abelian rings which are direct sums of fields, will be used to make certain there are enough minimal idempotents in a finite-dimensional, formally real algebra. The following lemmas will then lead us up to a point where the coordinatization theorem can be used.

2.9.2. Proposition. *Suppose $R=\mathbb{R}$, $\mathbb{C}$, $\mathbb{H}$ or $\mathbb{O}$. Let $n\geqslant 2$; if $R=\mathbb{O}$, assume $n\leqslant 3$. Then $H_n(R)$ is a formally real Jordan algebra.*

Proof. By 2.7.5, $H_3(\mathbb{O})$ is a Jordan algebra. Since $H_2(\mathbb{O})\simeq U_{e_{11}+e_{22}}H_3(\mathbb{O})$, $H_2(\mathbb{O})$ is a Jordan algebra as well. That the other examples mentioned are Jordan algebras is clear. (The algebras $H_2(R)$ are examples of spin factors, described in Chapter 6; see also 2.9.7.)

Let Tr be the usual trace in $H_n(R)$, i.e. $\mathrm{Tr}(a)=\sum_{i=1}^n a_{ii}$. Then, if $a\in H_n(R)$,

$$\mathrm{Tr}(a^2)=\sum_{i,j}\bar{a}_{ji}a_{ji}.$$

However, $\bar{a}_{ji}a_{ji}$ is a positive real number if $a_{ij}\neq 0$, so $\mathrm{Tr}(a^2)>0$ if $a\neq 0$. From this the formal reality immediately follows.

2.9.3. Lemma. *Let R be an Abelian ring without nilpotent elements such that any descending chain $I_1\supset I_2\supset\ldots$ of ideals in R is finite. Then there* *exist minimal idempotents $e_1, \ldots, e_n$ in R such that $e_ie_j = 0$ whenever $i \neq j$, $e_1 + \ldots + e_n$ is an identity in R, and Re_i is a field for $i = 1, \ldots, n$.*

Proof. Pick any nonzero minimal ideal I of R. It exists by virtue of the assumed 'descending chain condition'. We claim that I contains a nonzero idempotent e. Indeed, let x be a nonzero element of I. Because R has no nilpotents, $x^2 \neq 0$. But then $x^2 \in Ix$, and so Ix is a nonzero ideal contained in I. By the minimality of I, $Ix = I$. In particular $ex = x$ for some $e \in I$. Now we have $(e^2 - e)x = 0$, so that $e^2 - e$ belongs to the annihilator ideal in I of x. Since this ideal does not contain x it must be zero by the minimality of I. Thus $e^2 - e = 0$, so e is an idempotent.

Actually e is a unit for I, as we shall see next. For if $x \in I$ is a general element, then $e(x - ex) = 0$, so an argument with annihilators, like above, proves that $x - ex = 0$. Even more is true: I is a field, by the previous paragraph, for if $x \neq 0$ belongs to I then $e \in I = Ix$, so x has an inverse in I.

Let J be the annihilator in R of e. Then J is an ideal of R, and $R = I \oplus J$. Indeed, if $x \in R$ then $x = ex + (x - ex) \in I + J$, while if $x \in I \cap J$ then $x = ex = 0$.

If $J = 0$ then R is a field, and we are through. Otherwise, noting that J satisfies the same requirements we made for R, we can repeat the above argument with J replacing R. Thus, we get a sequence $I_1, I_2, \ldots$ of minimal ideals, which are fields with identities $e_1, e_2, \ldots,$ and ideals $J_1, J_2, \ldots$ such that $R = I_1 \oplus J_1$, $J_n = I_{n+1} \oplus J_{n+1}$. But then $J_1 \supset J_2 \supset \ldots$ is a descending chain of ideals, which by assumption must stop. This completes the proof, since now $I_k = Re_k$.

2.9.4 Lemma. *Let A be a finite-dimensional, formally real, unital Jordan algebra over $\mathbb{R}$. Then we have:*

(i) *A contains no nilpotent elements.*

(ii) *An idempotent p in A is minimal if and only if $\{pAp\} = \mathbb{R}p$.*

(iii) *Any element of A is contained in some maximal associative subalgebra of A, and every such subalgebra is of the form $\mathbb{R}p_1 \oplus \ldots \oplus \mathbb{R}p_n$, where $p_1, \ldots, p_n$ are pairwise orthogonal minimal idempotents with sum 1.*

(iv) *If p and q are orthogonal minimal idempotents in A and $a \in \{pAq\}$, then $a^2 = \lambda(p + q)$, where $\lambda \in \mathbb{R}$ and $\lambda \geq 0$. In particular, either $\{pAq\} = 0$ or p, q are strongly connected.*

(v) *If $a \in A$ and p is a minimal idempotent in A then $\{pa^2p\} = \lambda p$ where $\lambda \in \mathbb{R}$, $\lambda \geq 0$.*

(vi) *Let $a = \sum_{i=1}^n \alpha_i p_i$, where $p_1, \ldots, p_n$ are pairwise orthogonal minimal idempotents and $\alpha_i \in \mathbb{R}$. Then a is a square if and only if $\alpha_i \geq 0$, $i = 1, \ldots, n$.*

(vii) *If $a_1, \ldots, a_m \in A$ then $\sum_{j=1}^m a_j^2$ is a square.*

Proof. (i) If $a \in A$ is nilpotent, assume $a^{m-1} \neq 0$, $a^m = 0$. Let k be the integer satisfying $\frac{1}{2}m \leq k \leq \frac{1}{2}(m+1)$. Then $2k \geq m$, hence $a^{2k} = 0$. But then, by formal reality $a^k = 0$. Since $k < m$ this is a contradiction.

(ii) Clearly, if p and q are idempotents and $p \circ q = q$, then $q = \{pqp\} \in \{pAp\}$. Hence if $\{pAp\} = \mathbb{R}p$ we must have $q = 0$ or $q = p$, so that p is minimal. Conversely, assume that p is minimal, and let $a \in \{pAp\}$. Since the algebra generated by a and p satisfies the conditions of 2.9.3, it is a direct sum of fields, and in particular the identity p of this algebra is a sum of the identities $q_1, \ldots, q_n$ of these fields. By the minimality of p we must have $n = 1$, so the algebra generated by a and p is a field. The only finite-dimensional fields over $\mathbb{R}$ are $\mathbb{R}$ and $\mathbb{C}$ however (2.2.6), and the latter must be eliminated because it is not formally real. Hence this field is $\mathbb{R}p$, so $a \in \mathbb{R}p$, or $\{pAp\} = \mathbb{R}p$.

(iii) By power associativity (2.4.5) every element of A is contained in an associative subalgebra of A. Clearly, the union of an ascending chain of associative algebras is an associative algebra, so induction proves that there must be a maximal one containing the given element. (By Zorn's lemma this is true even in infinite dimensions.) Let R be a maximal associative subalgebra of A. Then $1 \in R$. By 2.9.3 there are idempotents $p_1, \ldots, p_n$ in R which are pairwise orthogonal and with sum 1, such that Rp_i is a field. Again, since this field is finite-dimensional and formally real over $\mathbb{R}$, it must be $\mathbb{R}p_i$, so that $R = \mathbb{R}p_1 \oplus \ldots \oplus \mathbb{R}p_n$. If p_1 is not minimal in A choose an idempotent q such that $p_1 \circ q = q$ while $0 \neq q \neq p_1$. Then $q = \{p_1qp_1\}$, so that $p_i \circ q = 0$ whenever $i \neq 1$. It follows that $\mathbb{R}q \oplus \mathbb{R}(p_1 - q) \oplus \mathbb{R}p_2 \oplus \ldots \oplus \mathbb{R}p_n$ is an associative algebra properly containing R, which is impossible. Hence p_1 is minimal, and similarly so are $p_2, \ldots, p_n$.

(iv) Let p, q be orthogonal minimal idempotents and $a \in \{pAq\}$. By (ii) and the multiplication rules for Peirce components, $a^2 = \lambda_1 p + \lambda_2 q$ where $\lambda_i \in \mathbb{R}$. This can be rewritten as $a^2 = \lambda(p+q) + \mu p$. However, a operator commutes with $p+q$ as well as with a^2, so if $\mu \neq 0$, then a operator commutes with p. Since $a \in \{pAq\}$, this implies $a = 0$. Therefore $\mu = 0$, so $a^2 = \lambda(p+q)$. If $\lambda < 0$ then $a^2 + [(-\lambda)^{1/2}(p+q)]^2 = 0$, which contradicts formal reality. Therefore, $\lambda \geq 0$. If $a \neq 0$ then $\lambda > 0$, so that $(\lambda^{-1/2}a)^2 = p+q$, whence p and q are strongly connected.

(v) Let $a \in A$, and let p be a minimal idempotent. By (iii) there are orthogonal minimal idempotents $p_1, \ldots, p_n$ with sum 1 and $p \in \mathbb{R}p_1 \oplus \ldots \oplus \mathbb{R}p_n$. Clearly $p = p_i$ for some i, say $p = p_1$. Write $a = \sum_{i \leq j} a_{ij}$, where $a_{ij} \in \{p_iAp_j\}$. From the multiplication rules for Peirce components we get

$$\{p_1a^2p_1\} = \sum_{j=1}^{n} \{p_1a_{1j}^2p_1\}.$$

Since $a_{11} \in \mathbb{R}p_1$, $a_{11}^2 = \lambda_1 p_1$, $\lambda_1 \geq 0$. By (iv) we also get $\{p_1a_{1j}^2p_1\} = \lambda_j p_1$, $\lambda_j \geq 0$, for $j = 2, \ldots, n$. Thus $\{pa^2p\} = \sum \lambda_j p$, and $\sum \lambda_j \geq 0$.

(vi) Let $a=\sum_{i=1}^{n}\alpha_i p_i$, where $p_1,\ldots,p_n$ are pairwise orthogonal minimal idempotents and $\alpha_i\in\mathbb{R}$. If $\alpha_i\geqslant 0$ then a is the square of $\sum_{i=1}^{n}\alpha_i^{1/2}p_i$. Conversely, if $a=b^2$, then $\alpha_i p_i=\{p_i b^2 p_i\}$ so $\alpha_i\geqslant 0$ by (v).

(vii) Let $a_1,\ldots,a_m\in A$ and let $a=\sum_{j=1}^{m}a_j^2$. Write $a=\sum\alpha_i p_i$, where $p_1,\ldots,p_n$ are pairwise orthogonal, minimal idempotents. Then

$$\alpha_i p_i=\{p_i a p_i\}=\sum_{j=1}^{m}\{p_i a_j^2 p_i\},$$

so $\alpha_i\geqslant 0$ by (v). By (vi) a is a square.

2.9.5. Lemma. *Let A be a finite-dimensional, unital, formally real Jordan algebra. Let $p_1,\ldots,p_n$ be a family of pairwise orthogonal minimal idempotents with sum* 1. *Write $p_i\sim p_j$ if p_i and p_j are strongly connected. Then $\sim$ is an equivalence relation, and for all i, $\sum_{p_j\sim p_i}p_j$ is a central idempotent in A.*

Proof. Clearly the relation $\sim$ is reflexive and symmetric. Transitivity follows from 2.8.20, so $\sim$ is an equivalence relation.

Let $e=\sum_{p_j\sim p_i}p_j$. To show that e is central it is enough to show that e operator commutes with any element of $A_{kl}=\{p_k A p_l\}$ for any k, l. This is clear if $p_k\sim p_i\sim p_l$, or if neither $p_k\sim p_i$ nor $p_l\sim p_i$, for then, if $a\in A_{kl}$, $a\circ e=a$ or $a\circ e=0$, so 2.5.5 shows $[T_a,T_e]=0$. If, on the other hand, $p_k\sim p_i\nsim p_l$, then $A_{kl}=0$, by 2.9.4(iv). This completes the proof.

2.9.6. Theorem. *Every finite-dimensional, formally real, unital Jordan algebra A is a direct sum of simple algebras. If A is simple then it contains $n\geqslant 1$ pairwise orthogonal and strongly connected minimal idempotents with sum* 1. *If $n=1$, $A\simeq\mathbb{R}$. If $n\geqslant 3$, A is isomorphic to one of $H_n(\mathbb{R})$, $H_n(\mathbb{C})$, $H_n(\mathbb{H})$ or, if $n=3$, $H_3(\mathbb{O})$.*

Proof. Let A be a finite-dimensional, formally real, unital Jordan algebra. Then by 2.9.4 A contains $n\geqslant 1$ pairwise orthogonal minimal idempotents $p_1,\ldots,p_n$ with sum 1. If e is a central projection in A then eA is an ideal, and $A=eA+(1-e)A$ is an algebra direct sum. Therefore, from 2.9.5 we may conclude two things. First, if A is simple there can be no nontrivial central projections, and so $p_1,\ldots,p_n$ are all pairwise strongly connected. Secondly, A is a direct sum of algebras which do contain pairwise orthogonal and strongly connected minimal idempotents with sum 1. If A satisfies this requirement we must show that A is simple. So let I be an ideal in A, let $a\in I$ be nonzero, and write $a=\sum_{i\leqslant j}a_{ij}$, where $a_{ij}\in A_{ij}=\{p_i A p_j\}$. Pick i, j so that $a_{ij}\neq 0$. If $i=j$, then $a_{ii}=\{p_i a p_i\}\in I$. Since by 2.9.4(ii) $a_{ii}=\lambda p_i$ with $\lambda\neq 0$, $p_i\in I$. If $i\neq j$ then by 2.9.4(iv) $a_{ij}^2=\lambda(p_i+p_j)$, $a_{ij}=2\{p_i a p_j\}\in I$, so again $p_i=\lambda^{-1}p_i\circ a_{ij}^2$ belongs to I. This implies that any $b\in A_{ki}$ belongs to I, for any k, since $b=2p_i\circ b$ (or

$b = p_i \circ b$, if $k = i$). Repeat the previous argument with any nonzero element of A_{ki} to conclude $p_k \in I$, so $1 = \sum p_k \in I$, and thus $I = A$, i.e. A is simple.

If $n = 1$ $A \simeq \mathbb{R}$ by 2.9.4(ii). If $n \geq 3$ the coordinatization theorem (2.8.9) tells us that $A \simeq H_n(R)$, for some real $*$ algebra R. Since $U_{e_{11}}(H_n(R)) = R_{sa}e_{11}$, 2.9.4(ii) implies that $R_{sa} = \mathbb{R}1$. Suppose that $x \in R$ is not zero. Let $a = xe_{12} + x^*e_{21}$. Then $0 \neq a^2 = (xx^*)e_{11} + (x^*x)e_{22}$. Thus either xx^* or x^*x is nonzero.

However, since $x + x^* \in R_{sa} = \mathbb{R}1$, x and x^* commute, so $x^*x \neq 0$. By 2.7.8 R is $*$ isomorphic to either $\mathbb{R}$, $\mathbb{C}$, $\mathbb{H}$ or, if $n = 3$, to $\mathbb{O}$.

2.9.7. To complete the classification of finite-dimensional, formally real Jordan algebras we must consider the case $n = 2$, left out of 2.9.6. This produces the so-called *spin factors*, which are radically different from other examples we have encountered so far. More precisely, a *finite-dimensional spin factor* is an algebra constructed in the following way. Start with a real vector space H of finite dimension ≥ 2. Assume that a positive definite symmetric inner product (,) is given on H. Then define the product in $H \oplus \mathbb{R}1$ by

$$(\xi \oplus \lambda 1) \circ (\eta \oplus \mu 1) = (\lambda \eta + \mu \xi) \oplus ((\xi, \eta) + \lambda \mu) 1.$$

This procedure may be described as 'appending a unit to H'. The reader may check that $H \oplus \mathbb{R}1$ is a formally real Jordan algebra, that it is also simple, and that, if ξ is a unit vector in H, then $\frac{1}{2}(1 + \xi)$ and $\frac{1}{2}(1 - \xi)$ are minimal projections with sum 1. It is also true, but not so obvious, that $H \oplus \mathbb{R}1$ is a special Jordan algebra. This may be proved in many different ways, e.g. using Clifford algebras. However, we shall return to all this in Chapter 6.

2.9.8. Theorem. *Any finite-dimensional, formally real, unital Jordan algebra which is also simple and contains two minimal projections with sum 1 is a finite-dimensional spin factor.*

Proof. Let p and q be minimal projections in such an algebra A, with $p + q = 1$. Clearly, p and q are orthogonal. By 2.9.4(ii), this gives rise to a Peirce decomposition

$$A = \mathbb{R}p \oplus \mathbb{R}q \oplus A_{12},$$

where $A_{12} = \{pAq\}$. By 2.9.4(iv), if $a \in A_{12}$ then $a^2 = \lambda 1$, where $\lambda \in \mathbb{R}$, $\lambda \geq 0$. Since $a \circ b = \frac{1}{2}[(a + b)^2 - a^2 - b^2]$, we have $a \circ b \in \mathbb{R}1$ for all $a, b \in A_{12}$. Thus, there is a symmetric, positive definite inner product on A_{12} such that

$$a \circ b = (a, b)1 \qquad (a, b \in A_{12}). \tag{2.98}$$

If $a \in A_{12}$ then $p \circ a = \frac{1}{2}a = q \circ a$, so that $(p-q) \circ a = 0$. Also, $(p-q)^2 = 1$. Therefore the above inner product can be extended to $H = \mathbb{R}(p-q) \oplus A_{12}$, so that (2.98) holds for $a, b \in H$. Note that $\dim H \geqslant 2$, for if $A_{12} = 0$ then $A = \mathbb{R}p \oplus \mathbb{R}q$ is not simple. Since $A = H \oplus \mathbb{R}1$, this completes the proof.

2.9.9. Remark. The reader may verify directly that the algebras $H_2(\mathbb{R})$, $H_2(\mathbb{C})$, $H_2(\mathbb{H})$ and $H_2(\mathbb{O})$ are finite-dimensional spin factors. Indeed, an orthonormal basis for the inner product space H consists of the first two (resp. three, five or nine) matrices of the following list:

$$\begin{pmatrix} 1 & 0 \\ 0 & -1 \end{pmatrix}, \quad \begin{pmatrix} 0 & 1 \\ 1 & 0 \end{pmatrix}, \quad \begin{pmatrix} 0 & i \\ -i & 0 \end{pmatrix},$$

$$\begin{pmatrix} 0 & j \\ -j & 0 \end{pmatrix}, \quad \begin{pmatrix} 0 & k \\ -k & 0 \end{pmatrix}, \quad \begin{pmatrix} 0 & l \\ -l & 0 \end{pmatrix},$$

$$\begin{pmatrix} 0 & il \\ -il & 0 \end{pmatrix}, \quad \begin{pmatrix} 0 & jl \\ -jl & 0 \end{pmatrix}, \quad \begin{pmatrix} 0 & kl \\ -kl & 0 \end{pmatrix}.$$

2.10. Comments

The topics presented in this chapter are by now quite classical, and have already appeared in book form [4, 9, 11]. Our aim has been to prove exactly what will be needed in the subsequent chapters and to classify the finite-dimensional formally real (unital) Jordan algebras. We have used several sources. In Section 2.2 we have followed Schafer [11]. Section 2.3 on special Jordan algebras is mostly extracted from Jacobson's book [9]. Our proof of Macdonald's theorem (2.4.13) is based on those of Jacobson [62] and McCrimmon [80]. In Section 2.7 our proof that $H_3(\mathbb{O})$ is a Jordan algebra is taken from Jacobson [9]. The coordinatization theorem (2.8.9) is due to Jacobson. Our treatment is based on his proof [9]. The classification of finite-dimensional simple formally real Jordan algebras is due to Jordan, von Neumann and Wigner [69]. Our proof is more in the spirit of the subsequent chapters in this book.

3

JB algebras

3.1. Definition of JB algebras

3.1.1. One of the major contributions to the theory of Jordan algebras was made by Jordan, von Neumann and Wigner [69] when they classified all simple finite-dimensional formally real Jordan algebras (theorem 2.9.6). We shall now introduce the proper Banach algebra version of the formally real Jordan algebras and shall consider a concrete case first. Our nomenclature will be a Jordanization of the corresponding usage in C^* algebras. Thus the reader will encounter the following concepts: JB algebras will correspond to (abstract) B^* algebras, JC algebras to (concrete) C^* algebras, JW and JBW algebras to W^* algebras, i.e. von Neumann algebras, in concrete and abstract formulations respectively.

3.1.2. Let H be a complex Hilbert space and $B(H)$ the algebra of all bounded linear operators on H. Let $B(H)_{\mathrm{sa}}$ denote the special Jordan algebra of all self-adjoint operators in $B(H)$ equipped with the operator norm. By a *JC algebra* we shall mean any norm-closed Jordan subalgebra of $B(H)_{\mathrm{sa}}$. We shall sometimes call a normed Jordan algebra a JC algebra if it is isometrically isomorphic to a JC algebra as defined above.

In finite dimensions $H_n(\mathbb{R})$, $H_n(\mathbb{C})$ and $H_n(\mathbb{H})$ are all JC algebras. Indeed, $H_n(\mathbb{C})=B(\mathbb{C}^n)_{\mathrm{sa}}$, where $\mathbb{C}^n$ is the n-dimensional Hilbert space, so $H_n(\mathbb{C})$ is a JC algebra by definition. $H_n(\mathbb{R})$ is a norm-closed Jordan subalgebra of $H_n(\mathbb{C})$, hence is a JC algebra. Since $\mathbb{H}$ can be identified with a real $*$ subalgebra of $M_2(\mathbb{C})$ via the representation

$$a+bi+cj+dk \to \begin{pmatrix} a+bi & c-di \\ -c-di & a-bi \end{pmatrix}, \qquad a, b, c, d \in \mathbb{R},$$

$H_n(\mathbb{H})$ is a Jordan subalgebra of $H_{2n}(\mathbb{C})$, hence is also a JC algebra.

Note that by 2.8.5 $H_3(\mathbb{O})$ is not a JC algebra.

3.1.3. A *Jordan Banach algebra* is a real Jordan algebra A equipped with a complete norm satisfying

$$\|a \circ b\| \leq \|a\| \, \|b\|, \qquad a, b \in A.$$

3.1.4. A *JB algebra* is a Jordan Banach algebra A in which the norm satisfies the following two additional conditions for $a, b \in A$:
(i) $\|a^2\| = \|a\|^2$,
(ii) $\|a^2\| \leq \|a^2 + b^2\|$.
If A is unital we denote the identity by 1. Then it is clear from (i) that $\|1\| = 1$.

It is clear from 1.3.2 that each JC algebra is a JB algebra, and we shall see below that $H_3(\mathbb{O})$ is also a JB algebra, and indeed all finite-dimensional (unital) formally real Jordan algebras are JB algebras. We note, however, that this is false in infinite dimensions. Indeed, let A denote the set of complex functions $f(z)$, analytic for $|z| < 1$, continuous for $|z| = 1$ and real for z real. With pointwise multiplication and norm given by

$$\|f\| = \sup_{|z| \leq 1} |f(z)|,$$

A is a formally real Jordan Banach algebra satisfying (i) but not (ii). Pick, for example, f and g in A such that $f(z)^2 = z^2 - z^4$ and $g(z)^2 = z^4$. Then $\|f^2\| = 2$, but $\|f^2 + g^2\| = 1$.

3.1.5. An important aspect of JB algebras is their order structure. We shall show that unital JB algebras can be characterized among all real Jordan algebras as those which are order unit spaces (see 1.2.1) with certain properties. We begin by showing the first half of this characterization, and postpone the second half until later (3.3.10).

3.1.6. Proposition. *Suppose A is a complete order unit space which is a Jordan algebra for which the distinguished order unit acts as an identity element, and suppose*

$$-1 \leq a \leq 1 \quad \textit{implies} \quad 0 \leq a^2 \leq 1 \quad \textit{for } a \in A. \tag{3.1}$$

Then A is a JB algebra in the order norm.

Proof. Consider two elements a, b in A with $\|a\| \leq 1$, $\|b\| \leq 1$. Then $\|\frac{1}{2}(a+b)\| \leq 1$ and $\|\frac{1}{2}(a-b)\| \leq 1$, so that $-1 \leq \frac{1}{2}(a+b) \leq 1$ and $-1 \leq \frac{1}{2}(a-b) \leq 1$. By (3.1) $0 \leq [\frac{1}{2}(a+b)]^2 \leq 1$ and $0 \leq [\frac{1}{2}(a-b)]^2 \leq 1$, so that

$$-1 \leq \tfrac{1}{4}[(a+b)^2 - (a-b)^2] \leq 1.$$

It follows that

$$\|a \circ b\| = \tfrac{1}{4}\| (a+b)^2 - (a-b)^2 \| \leqslant 1,$$

proving that A is a Jordan Banach algebra.

Assume next $\|a^2\| \leqslant 1$. Then $a^2 \leqslant 1$ by definition of the order norm. Since by (3.1) all squares are positive we obtain

$$a = \tfrac{1}{2}[a^2 + 1 - (a-1)^2] \leqslant \tfrac{1}{2}(a^2+1) \leqslant 1$$

and

$$a = \tfrac{1}{2}[(a+1)^2 - a^2 - 1] \geqslant \tfrac{1}{2}(-a^2-1) \geqslant -1,$$

which gives $-1 \leqslant a \leqslant 1$, i.e. $\|a\| \leqslant 1$. Thus $\|a^2\| \leqslant 1$ implies $\|a\|^2 \leqslant 1$, and hence for all a, $\|a\|^2 \leqslant \|a^2\|$. Since A is a Jordan Banach algebra the converse inequality is obvious, so (i) follows.

By (3.1) sums of squares are positive, so if $a, b \in A$ then $0 \leqslant a^2 \leqslant a^2 + b^2$. It follows that

$$\begin{aligned} \|a^2\| &= \inf\{\lambda > 0 \colon 0 \leqslant a^2 \leqslant \lambda 1\} \\ &\leqslant \inf\{\lambda > 0 \colon 0 \leqslant a^2 + b^2 \leqslant \lambda 1\} \\ &= \|a^2 + b^2\|. \end{aligned}$$

This establishes (ii), so A is a JB algebra.

3.1.7. Corollary. *Every unital finite-dimensional formally real Jordan algebra A is a* JB *algebra.*

Proof. By 2.9.4 the set of squares from a proper cone in A, hence A is a partially ordered vector space. By the same lemma, if $a \in A$ there exist orthogonal idempotents $p_1, \ldots, p_m$ in A with sum 1 such that $a = \sum_{k=1}^m \alpha_k p_k$, $\alpha_k \in \mathbb{R}$. Moreover a is a square if and only if each $\alpha_k \geqslant 0$, from which it follows that 1 is an order unit. If

$$a \leqslant \frac{1}{n} 1 = \sum_{k=1}^{m} \frac{1}{n} p_k$$

for all $n \in \mathbb{N}$, then $\alpha_k \leqslant 1/n$ for all n, hence $\alpha_k \leqslant 0$ and $a \leqslant 0$. Thus A is Archimedean, and so is an order unit space. If $-1 \leqslant a \leqslant 1$ then $-1 \leqslant \alpha_k \leqslant 1$ for all k, hence $a^2 = \sum_{k=1}^m \alpha_k^2 p_k \leqslant 1$, so A is a JB algebra by 3.1.6.

3.1.8. Remark. It was noticed in 3.1.2 that the algebras $H_n(\mathbb{R})$, $H_n(\mathbb{C})$ and $H_n(\mathbb{H})$ are all JC algebras. In Chapter 6 we shall see that the finite-dimensional spin factors (2.9.7) are JC algebras too. By 2.9.6 and 2.9.8 the only formally real finite-dimensional Jordan algebra remaining is $H_3(\mathbb{O})$. It was noted in 3.1.2 that $H_3(\mathbb{O})$ is not a JC algebra; however, by 3.1.7 it is a JB algebra.

3.2. Spectral theory

3.2.1. Just as for C^* algebras spectral theory will be a major technical tool in the theory of JB algebras. We shall prove two spectral theorems, one for general associative JB algebras and one for singly generated ones. Among all the available proofs we shall choose the quickest, namely to reduce to the well known complex case (1.3.3 and 1.3.4).

If X is a locally compact Hausdorff space then from 1.3.2 $C_0(X)$ is the self-adjoint part of the Abelian C^* algebra $C_0^{\mathbb{C}}(X)$. Thus $C_0(X)$ is an associative JC algebra under pointwise multiplication and supremum norm. We next show the converse result.

3.2.2. Theorem. *Let A be an associative* JB *algebra. Then there is a locally compact Hausdorff space X such that A is isometrically isomorphic to $C_0(X)$. Furthermore, A is unital if and only if X is compact.*

Proof. Let $\mathscr{A}$ denote the complexification of A, i.e. $\mathscr{A}=\{a+ib\colon a, b\in A\}$. With the product

$$(a+ib)(c+id)=(a\circ c-b\circ d)+i(a\circ d+b\circ c)$$

and involution $(a+ib)^*=a-ib$, $\mathscr{A}$ becomes an involutive Abelian complex algebra. Define

$$\|a+ib\|=\|a^2+b^2\|^{1/2}, \qquad a, b\in A. \tag{3.2}$$

We shall show that $\|\ \ \|$ is a norm on $\mathscr{A}$ making $\mathscr{A}$ into an Abelian C^* algebra containing A as its self-adjoint part. By the characterization of Abelian C^* algebras (1.3.3) this will prove the theorem.

Let $z=a+ib\in\mathscr{A}$. Then $\|z\|^2=\|z^*z\|$, so that $\|\ \ \|$ is a C^* norm if it is a Banach algebra norm. Clearly $\|\lambda z\|=|\lambda|\,\|z\|$ if $\lambda\in\mathbb{C}$, and if $w\in\mathscr{A}$ then by the C^* property of $\|\ \ \|$,

$$\|zw\|^2=\|w^*z^*zw\|=\|(z^*z)\circ(w^*w)\|\leqslant\|z^*z\|\,\|w^*w\|=\|z\|^2\,\|w\|^2,$$

which shows $\|zw\|\leqslant\|z\|\,\|w\|$ for all $z, w\in\mathscr{A}$.

We next show the triangle inequality. Let $z=a+ib$, $w=c+id$ belong to $\mathscr{A}$. Then we have

$$z^*w+w^*z=2a\circ c+2b\circ d\in A. \tag{3.3}$$

Furthermore, by the defining properties of JB algebras and associativity of A we find

$$\begin{aligned}\|a\circ c+b\circ d\|^2 &\leqslant \|(a\circ c+b\circ d)^2+(a\circ d-b\circ c)^2\| \\ &= \|(a^2+b^2)\circ(c^2+d^2)\| \\ &\leqslant \|a^2+b^2\|\,\|c^2+d^2\| \\ &= \|z\|^2\,\|w\|^2.\end{aligned}$$

Therefore, by (3.3) we have

$$\begin{aligned}\|z+w\|^2 &= \|(z+w)^*(z+w)\| \\ &\leqslant \|z^*z\|+\|z^*w+w^*z\|+\|w^*w\| \\ &\leqslant \|z\|^2+2\,\|z\|\,\|w\|+\|w\|^2 \\ &= (\|z\|+\|w\|)^2,\end{aligned}$$

proving the triangle inequality. Since it is clear from (3.2) that $\mathscr{A}$ is complete, we have shown that $\mathscr{A}$ is a C^* algebra, proving the theorem.

3.2.3. If A is a unital Jordan Banach algebra and $a\in A$ we denote by $C(a)$ the smallest norm-closed Jordan subalgebra of A containing a and 1. Then $C(a)$ is associative. We define the *spectrum of a*, denoted by Sp a, to be the set of $\lambda\in\mathbb{R}$ such that $a-\lambda 1$ does not have an inverse in $C(a)$.

3.2.4. The spectral theorem. *Let A be a unital Jordan Banach algebra. Let $a\in A$ and suppose $C(a)$ is a* JB *algebra in an equivalent norm* $|||\ \ |||$. *Then $C(a)$ is isometrically isomorphic to $C(\mathrm{Sp}\ a)$ with respect to* $|||\ \ |||$.

Proof. By 3.2.2 and its proof there is a compact Hausdorff space X such that $C(a)$ with the norm $|||\ \ |||$ is mapped isometrically and isomorphically onto $C(X)$. Thus Sp a as defined above equals the spectrum of a as an element in $C(X)$. Thus by the complex spectral theorem (1.3.4) $C(a)$ is isometrically isomorphic to $C(\mathrm{Sp}\ a)$.

3.2.5. Corollary. *Let A and B be unital* JB *algebras and π a unital homomorphism of A into B. Then for each $a\in A$, $C(\pi(a))=\pi(C(a))$.*

Proof. If $a\in A$ is invertible then clearly $\pi(a)$ is invertible. It follows that $\mathrm{Sp}\ \pi(a)\subseteq \mathrm{Sp}(a)$. If $f\in C(\mathrm{Sp}\ a)$ recall from 1.3.5 the definition of $f(a)$ through the isomorphism $C(a)\cong C(\mathrm{Sp}\ a)$. Then $\pi(f(a))=f(\pi(a))$. Indeed, this is trivial if f is a polynomial, and follows in general from the Stone–Weierstrass theorem [7, IV.6.16]. The corollary follows from 3.2.4 and the Tietze extension theorem [7, I.5.3].

3.2.6. Proposition. *Let A be a unital Jordan Banach algebra such that*

$C(a)$ is a JB algebra in the given norm for each $a \in A$. Then A is a JB algebra.

Proof. Let $S(A) = \{\rho \in A^*: \|\rho\| = \rho(1)\}$, and let $A^+ = \{a \in A: \rho(a) \geqslant 0$ for all $\rho \in S(A)\}$. Then clearly A^+ is a cone. By the spectral theorem (3.2.4) and the fact that $C(\mathrm{Sp}\, a)$ is an order unit space with the identity as order unit and the norm being the same as the order norm, we have by 1.2.5 that $a \geqslant 0$ as an element in $C(a)$ if and only if $\rho(a) \geqslant 0$ for all states ρ of $C(a)$. By the Hahn–Banach theorem (1.1.12) and 1.2.2 each state of $C(a)$ has an extension to a functional in $S(A)$, and each functional in $S(A)$ restricts to a state on $C(a)$. Thus the positive part of $C(a)$ is $C(a) \cap A^+$, so A^+ is a proper cone with respect to which A is an order unit space in which the norm coincides with the given norm. Since $-1 \leqslant a \leqslant 1$ implies $a^2 \leqslant 1$ in $C(a)$ so in A for each $a \in A$, A is a JB algebra by 3.1.6.

3.2.7. Recall that if a and b belong to a Jordan algebra, then $U_a b = \{aba\} = 2a \circ (a \circ b) - a^2 \circ b$. Hence U_a is a continuous operator on a Jordan Banach algebra for each element a, and $\|U_a\| \leqslant 3\|a\|^2$.

3.2.8. Lemma. *Let A be a unital Jordan Banach algebra such that $C(a)$ is isomorphic to a JB algebra for each $a \in A$. Let $a \in A$ and $\lambda \in \mathbb{R}$. Then $\lambda \in \mathrm{Sp}\, a$ if and only if $U_{\lambda 1 - a}$ has no bounded inverse as an operator on A.*

Proof. Let b be an element in A such that U_b has a bounded inverse. Then there exists $k > 0$ $(k = \|U_b^{-1}\|^{-1})$ such that $\|U_b x\| \geqslant k\|x\|$ for all $x \in A$. If b is not invertible in $C(b)$ there is by the spectral theorem (3.2.4) a sequence (x_n) in $C(b)$ with $\|x_n\| = 1$ such that $\|U_b x_n\| \to 0$, contradicting the above inequality. Thus b is invertible. Conversely, if b is invertible in $C(b)$ let c be its inverse in $C(b)$. From the identity $1 = U_1 = U_{\{bc^2b\}} = U_b U_{c^2} U_b$ (2.4.18) U_b has a bounded inverse, completing the proof of the lemma.

3.2.9. If A is as in 3.2.8 we define the *inverse* of an element $a \in A$ to be its inverse in $C(a)$, if it exists, and denote it by a^{-1}. It should be remarked that this definition is equivalent to the usual definition of inverse in Jordan algebras; namely, b is the inverse of a if b satisfies $a \circ b = 1$ and $a^2 \circ b = a$. However, we shall not need this latter definition.

3.2.10. Lemma. *Let A be a unital Jordan Banach algebra for which $C(a)$ is isomorphic to a JB algebra for each $a \in A$. Suppose $a \in A$ is invertible. Then U_a is invertible as an operator on A, and $(U_a)^{-1} = U_{a^{-1}}$.*

Proof. It was shown in the proof of 3.2.8 that U_a is invertible. From the identity $U_{\{aba\}} = U_a U_b U_a$ (2.4.18) we have $U_a = U_a U_{a^{-1}} U_a$, hence $1 = (U_a)^{-1} U_a U_{a^{-1}} U_a = U_{a^{-1}} U_a$, proving the lemma.

3.2.11. Corollary. *If a and b are invertible elements in a unital* JB *algebra then so is $\{aba\}$, with inverse $\{a^{-1}b^{-1}a^{-1}\}$.*

Proof. By 3.2.8 and 2.4.18 $U_{\{aba\}} = U_a U_b U_a$ is invertible, and so $\{aba\}$ is invertible. By 3.2.10 and 2.4.18 we get

$$\begin{aligned}\{aba\}^{-1} &= U_{\{aba\}^{-1}}\{aba\} = U_{\{aba\}}^{-1}\{aba\} \\ &= U_a^{-1}U_b^{-1}U_a^{-1}U_a b = \{a^{-1}b^{-1}a^{-1}\}.\end{aligned}$$

3.3. Order structure in JB algebras

3.3.1. We saw in 3.1.6 that a Jordan algebra which is also an order unit space with certain properties is automatically a JB algebra. This result together with experience from C^* algebras indicates that the order structure is very important in JB algebras. We shall in the present section prove the basic results along these lines, one of which is the converse to 3.1.6 for unital JB algebras. For non-unital JB algebras the situation is not so clear, but can be resolved by adjoining an identity. While this is easy for C^* algebras, it is not so simple to prove that we get a JB algebra when we adjoin a unit to a non-unital JB algebra.

Let A be a JB algebra. We let $\tilde{A} = A \oplus \mathbb{R}$ with pointwise addition, multiplication given by

$$(a, \lambda) \circ (b, \mu) = (a \circ b + \mu a + \lambda b, \lambda\mu), \qquad a, b \in A, \quad \lambda, \mu \in \mathbb{R},$$

and norm given by

$$\|(a, \lambda)\|_1 = \|a\| + |\lambda|.$$

By 2.4.2 we know $\tilde{A}$ is a Jordan algebra. It is clear that $\|\ \ \|_1$ is a norm and that multiplication satisfies

$$\|x \circ y\|_1 \leq \|x\|_1 \|y\|_1, \qquad x, y \in \tilde{A}. \tag{3.4}$$

Let B be a unital closed subalgebra of $\tilde{A}$, and consider A as a Jordan subalgebra of $\tilde{A}$ via the identification of a and $(a, 0)$. Then we have $B = (B \cap A)^{\sim}$, so B is itself a JB algebra with the identity adjoined. In particular, if $a \in \tilde{A}$ then $C(a)$ is by 3.2.2 isometrically isomorphic to $C_0(X)^{\sim}$ for some locally compact Hausdorff space X. In particular $C(a)$ is isomorphic to some $C(Y)$ with Y a compact Hausdorff space, so it follows that $C(a)$ is a JB algebra in an equivalent norm. Thus $\tilde{A}$ is a Banach Jordan algebra for which $C(a)$ is isomorphic to a JB algebra for all $a \in \tilde{A}$.

3.3.2. With $\tilde{A}$ as above we define the *spectral radius norm* for an element $a \in \tilde{A}$ by

$$\|a\| = \sup\{|\lambda| : \lambda \in \operatorname{Sp} a\}.$$

From the previous paragraph it is clear that $\| \quad \|$ extends the norm on A; however, it is not clear that $\| \quad \|$ is a norm on $\tilde{A}$. Let $b \in A$ and choose $\mu \in \operatorname{Sp} b$ such that $|\mu| = \|b\|$. If $\lambda \in \mathbb{R}$, by the spectral theorem (3.2.4) it is easy to verify the following:

$$\|(b, \lambda)\| = \|b\| + |\lambda| \qquad \text{if } \lambda\mu \geqslant 0,$$

$$\|(b, \lambda)\| \geqslant \max\{\|b\| - |\lambda|, |\lambda|\} \qquad \text{if } \lambda\mu \leqslant 0.$$

Thus in every case $2\|(b, \lambda)\| \geqslant \|b\|$ and $\|(b, \lambda)\| \geqslant |\lambda|$. We therefore obtain the estimate

$$\|a\| \leqslant \|a\|_1 \leqslant 3\|a\| \qquad \text{for } a \in \tilde{A}. \tag{3.5}$$

3.3.3. Let A be a JB algebra. An element $a \in \tilde{A}$ is called *positive*, written $a \geqslant 0$, if $\operatorname{Sp} a \subset [0, \infty)$. By the spectral theorem (3.2.4) $a \geqslant 0$ if and only if a is a square. We write

$$\tilde{A}_+ = \{a \in \tilde{A} : a \geqslant 0\} = \{a^2 : a \in \tilde{A}\},$$

$$A_+ = \tilde{A}_+ \cap A = \{a^2 : a \in A\}.$$

3.3.4. Lemma. *Let A be a JB algebra. Suppose a is a positive invertible element in $\tilde{A}$. If $b \in \tilde{A}$ satisfies $\|b - a\|_1 < \frac{1}{27}\|a^{-1}\|^{-1}$ then b is invertible.*

Proof. By 3.2.7 U_a satisfies the estimate $\|U_a x\|_1 \leqslant 3\|a\|_1^2 \|x\|_1$ for $x \in \tilde{A}$. Apply this to the inverse of the positive square root $a^{1/2}$ of a, which exists in $C(a)$ by the spectral theorem (3.2.4). Then by (3.5) we have

$$\begin{aligned} \|U_{a^{-1/2}}(b-a)\|_1 &\leqslant 3\|a^{-1/2}\|_1^2 \|b-a\|_1 \\ &\leqslant 3 \times 9 \|a^{-1/2}\|^2 \|b-a\|_1 \\ &= 27\|a^{-1}\| \|b-a\|_1 \\ &< 1. \end{aligned}$$

Thus $1 + U_{a^{-1/2}}(b-a)$ is invertible in $\tilde{A}$, so by 3.2.11 $b = U_{a^{1/2}}(1 + U_{a^{-1/2}}(b-a))$ is invertible.

3.3.5. Lemma. *Let A be a JB algebra. Then $\tilde{A}_+$ is a closed subset of $\tilde{A}$.*

Proof. First we note that, if $a \in \tilde{A}_+$ and $\lambda \in (0, \infty)$, then $a + \lambda 1$ is invertible with $\|(a + \lambda 1)^{-1}\| \leqslant \lambda^{-1}$, where we write 1 for the unit $(0, 1)$ of $\tilde{A}$. Indeed, this is clear from the spectral theorem (3.2.4), since the restriction to $C(a)$ of $\| \quad \|$ is a norm making $C(a)$ into a JB algebra.

Suppose now that $a_n \in \tilde{A}_+$ and $\|a_n - a\|_1 \to 0$, and suppose further that a is nonpositive. Then we can find $\lambda > 0$ such that $a + \lambda 1$ is not invertible. By the above remark, however, $a_n + \lambda 1$ is invertible for all n, and $\|(a_n+\lambda 1)^{-1}\|\leqslant\lambda^{-1}$. When n is sufficiently large $\|(a+\lambda 1)-(a_n+\lambda 1)\|_1=\|a-a_n\|_1\leqslant\lambda/27$, and so by 3.3.4 $a+\lambda 1$ is invertible. This is a contradiction.

3.3.6. Proposition. *Let A be a JB algebra and $a, b\in\tilde{A}$. If $b\geqslant 0$ then $\{aba\}\geqslant 0$.*

Proof. We first assume that both a and b are invertible. Assume that $\{aba\}$ is not positive. For $t\in[0, 1]$ let

$$c_t=\{a((1-t)b+t1)a\}.$$

From the spectral theorem and (the proof of) 3.2.11 we see that c_t is invertible in $\tilde{A}$ for all t. Further, $c_0=\{aba\}$, $c_1=a^2$. By 3.2.7 the family (c_t) is also continuous and bounded.

Define the subset I of $[0, 1]$ by

$$I=\{t\colon c_t \text{ is nonpositive}\}.$$

Then $0\in I, 1\notin I$. Let $t_0=\sup\{t\colon t\in I\}$. If $t_0<1$ then, for all $t>t_0$, c_t is positive. Then by 3.3.5 and the continuity of (c_t), c_{t_0} is also positive. Of course, this is also true if $t_0=1$. Let $\varepsilon=\|c_{t_0}^{-1}\|^{-1}/27$. We shall prove that, whenever $\|c_t-c_{t_0}\|_1<\varepsilon$, c_t is positive. This will contradict the definition of t_0, and thereby complete the proof in this first case.

Therefore, assume that $\|c_t-c_{t_0}\|_1<\varepsilon$, and pick any $\lambda>0$. Since c_{t_0} is positive, the spectral theorem will imply that $\|(c_{t_0}+\lambda 1)^{-1}\|\leqslant\|c_{t_0}^{-1}\|$, so that

$$\|(c_{t_0}+\lambda 1)^{-1}\|^{-1}\geqslant 27\varepsilon.$$

Since $\|(c_t+\lambda 1)-(c_{t_0}+\lambda 1)\|_1=\|c_t-c_{t_0}\|_1<\varepsilon$, it follows from 3.3.4 that $c_t+\lambda 1$ is invertible. Because this holds for any $\lambda>0$, c_t is positive, as desired.

By the spectral theorem the invertible elements are dense in $\tilde{A}_+$. Since U_a is a continuous operator on $\tilde{A}$ it follows that $U_a b\geqslant 0$ whenever $b\geqslant 0$ and a is invertible.

Let now b be positive and invertible and a be general. Let $c=b^{1/2}$ be the positive square root of b in $C(b)$. From the identity $\{cac\}^2=\{c\{aba\}c\}$ (2.4.17) we have $U_cU_ab=(U_ca)^2\geqslant 0$. Since by 3.2.10 $U_{c^{-1}}=U_c^{-1}$, the previous paragraph applied to $U_{c^{-1}}$ shows $U_ab=U_{c^{-1}}(U_cU_ab)\geqslant 0$.

Finally, for general a and b with $b\geqslant 0$ we can again approximate b by positive invertible elements and use the continuity of U_a to conclude $U_ab\geqslant 0$.

3.3.7. Lemma. *Let A be a JB algebra. Then both $\tilde{A}_+$ and A_+ are proper convex cones.*

Proof. Since $A_+ = \tilde{A}_+ \cap A$ it suffices to show $\tilde{A}_+$ is a proper convex cone. It is clearly homogeneous. Let $a, b \in \tilde{A}$ and choose $\lambda > 0$. By the spectral theorem (3.2.4) $c = a^2 + \lambda 1$ is positive and invertible, hence has a positive invertible square root $c^{1/2}$. Then we have $a^2 + b^2 + \lambda 1 = U_{c^{1/2}}(1 + U_{c^{-1/2}} b^2)$, where by 3.3.6 $U_{c^{-1/2}} b^2 \geq 0$. Thus $1 + U_{c^{-1/2}} b^2$ is invertible, and so by 3.2.10 $a^2 + b^2 + \lambda 1$ is invertible. Therefore $a^2 + b^2 \in \tilde{A}_+$, which is therefore a convex cone. It is proper since $\mathrm{Sp}\, a \subset [0, \infty)$ for each $a \in \tilde{A}_+$.

3.3.8. Corollary. *A JB algebra is formally real.*

Proof. Suppose $\sum_{i=1}^n a_i^2 = 0$ for a_i in a JB algebra A, $i = 1, \ldots, n$. If $k \in \{1, \ldots, n\}$, since sums of squares are squares by 3.3.7, there is $b \in A$ such that $\sum_{i \neq k} a_i^2 = b^2$. Thus $\|a_k\|^2 = \|a_k^2\| \leq \|a_k^2 + b^2\| = 0$.

3.3.9. Theorem. *Let A be a JB algebra. Then the spectral radius norm is a norm on $\tilde{A}$ making $\tilde{A}$ into a JB algebra.*

Proof. By 3.3.7 $\tilde{A}_+$ is a proper cone. By the spectral theorem (3.2.4), for each $a \in \tilde{A}$, $C(a)$ is an order unit space with order unit 1 satisfying $-1 \leq b \leq 1$ implies $0 \leq b^2 \leq 1$, and in which the order norm equals the spectral radius norm. Thus $\tilde{A}$ is an order unit space with the same properties, hence the theorem follows from 3.1.6.

3.3.10. Proposition. *If A is a unital JB algebra, then A is a complete order unit space with the ordering induced by A_+ and order unit the identity 1. The order norm is the given one, and $a \in A$ satisfies $-1 \leq a \leq 1$ implies $0 \leq a^2 \leq 1$.*

Proof. The result, which is the converse to 3.1.6, is proved by an immediate modification of the proof of 3.3.9.

3.4 Ideals in JB algebras

3.4.1. Let A be a JB algebra and J a norm-closed ideal in A. Then J is also an ideal in $\tilde{A}$, as is easily seen from the definition of the product in $\tilde{A}$ (3.3.1). It is also easy to see that A/J is a Jordan algebra. Indeed, it is a Jordan Banach algebra in the quotient norm, for if $a, b \in A$ we have

$$\|a \circ b + J\| = \inf_{c \in J} \|a \circ b + c\| \leq \inf_{c,d \in J} \|(a+c) \circ (b+d)\|$$

$$\leq \inf_{c,d \in J} \|a+c\|\,\|b+d\| = \|a+J\|\,\|b+J\|.$$

3.4.2. Theorem. *Let A be a* JB *algebra and J a norm-closed ideal in A. Then A/J is a* JB *algebra in the quotient norm.*

Proof. Since by 3.4.1 J is an ideal in $\tilde{A}$, and $\tilde{A}/J=\widetilde{A/J}$, we may assume A is unital. Let $\pi: A\rightarrow A/J$ be the quotient map. By the spectral theorem (3.2.4) $C(a)$ is isometrically isomorphic to $C(\mathrm{Sp}\, a)$, and by 1.3.6 and 1.3.7 the image of a continuous homomorphism of $C(\mathrm{Sp}\, a)$ is itself the continuous real functions on a compact Hausdorff space. Thus $\pi(C(a))$ is a JB algebra in the quotient norm for each $a\in A$. Then the same is true for $C(\pi(a))=\pi(C(a))$, and so by 3.2.6 $A/J=\pi(A)$ is a JB algebra.

3.4.3. Proposition. *Let A and B be* JB *algebras and ϕ a homomorphism of A into B. Then $\|\phi(a)\|\leq\|a\|$ for all $a\in A$, and $\phi(A)$ is a* JB *subalgebra of B. Furthermore, if ϕ is injective, then it is an isometry.*

Proof. Considering $\tilde{A}$ and $\{\phi(1)\tilde{B}\phi(1)\}$ if necessary we may assume A, B and ϕ are all unital. Let $a\in A$. Then $\phi: C(a)\rightarrow C(\phi(a))$ is a homomorphism, hence $\|\phi(a)\|\leq\|a\|$ by the spectral theorem (3.2.4) and the same estimate for homomorphisms on $C(\mathrm{Sp}\, a)$ (1.3.7). Since ϕ is continuous its kernel J is a norm-closed ideal in A. Let π be the quotient map of A onto A/J and α the canonical isomorphism $\alpha: A/J\rightarrow\phi(A)$. Since α restricted to $\pi(C(a))$ is an isomorphism onto $C(\phi(a))$, and $\pi(C(a))=C(\pi(a))$ is a JB subalgebra of the JB algebra A/J (3.4.2), α is an isometry on $\pi(C(a))$ by 1.3.7. In particular α is an isometry on A/J. Since A/J is complete, so is $\phi(A)$, so that $\phi(A)$ is a JB subalgebra of B. Finally, if ϕ is injective then by 1.3.7 ϕ restricted to each $C(a)$ is isometric, hence ϕ is an isometry.

3.4.4. In the course of the proof of 3.4.3 we showed that $\|\phi(a)\|=\|a+J\|$—the quotient norm of $a+J$ in A/J.

3.5. Approximate identities

3.5.1. An *increasing approximate identity* in a JB algebra A is a family $(u_\lambda)_{\lambda\in J}$ of elements in A indexed by a directed set J such that

(i) $\|u_\lambda\|\leq 1$ for all $\lambda\in J$,

(ii) $0\leq u_\lambda\leq u_\mu$ whenever $\lambda\leq\mu$ in J,

(iii) $\|u_\lambda\circ a-a\|\rightarrow 0$ for all $a\in A$.

In C^* algebras such approximate identities are usually used in order to prove the results on ideals in the previous section. We shall also need them in order to study ideals, but in a different connection.

3.5.2. Lemma. *Let A be a JB algebra and $a, b \in A$. Then we have:*
(i) $\|\{ab^2a\}\| = \|\{ba^2b\}\|$,
(ii) $\|a \circ b\|^2 \leq \|a\| \, \|\{bab\}\|$ *for* $a \geq 0$.

Proof. We may by 3.3.9 assume A is unital. Then we have from the positivity of U_a and U_b (3.3.6), 2.4.17 and 2.4.21

$$\begin{aligned}\{ab^2a\}^2 &= \{a\{b\{ba^2b\}b\}a\} \\ &\leq \|\{ba^2b\}\| \, \{a\{b1b\}a\} \\ &= \|\{ba^2b\}\| \, \{ab^2a\}.\end{aligned}$$

Thus $\|\{ab^2a\}\|^2 \leq \|\{ba^2b\}\| \, \|\{ab^2a\}\|$, and (i) follows by cancellation and symmetry.

To show (ii) note that $(a \circ b)^2 = \frac{1}{2} a \circ \{bab\} + \frac{1}{4}\{ab^2a\} + \frac{1}{4}\{ba^2b\}$, as follows from Macdonald's theorem. Since $a \geq 0$, $a^2 \leq \|a\| \, a$, so by 3.3.6, $\{ba^2b\} \leq \|a\| \{bab\}$; thus (ii) is immediate from (i).

3.5.3. Lemma. *Let A be a unital JB algebra and $a, b \in A$. Suppose a is invertible and $0 \leq a \leq b$. Then b is invertible and $0 \leq b^{-1} \leq a^{-1}$.*

Proof. By 2.4.17 and 3.3.6 we have

$$\begin{aligned}\{b^{1/2}a^{-1}b^{1/2}\}^2 &= \{b^{1/2}\{a^{-1}ba^{-1}\}b^{1/2}\} \geq \{b^{1/2}\{a^{-1}aa^{-1}\}b^{1/2}\} \\ &= \{b^{1/2}a^{-1}b^{1/2}\} \geq 0.\end{aligned}$$

Since in general for $x \in A$, $x \geq 0$ and $x^2 \geq x \neq 0$ imply $x \geq 1$, we have $\{b^{1/2}a^{-1}b^{1/2}\} \geq 1$. Since a is positive and invertible there is by the spectral theorem (3.2.4) $\varepsilon > 0$ such that $\varepsilon 1 \leq a$. But then $\varepsilon 1 \leq b$, so b is invertible. Since by 3.2.10 $U_{b^{-1/2}} = (U_{b^{1/2}})^{-1}$ we have, using 3.3.6,

$$a^{-1} = \{b^{-1/2}\{b^{1/2}a^{-1}b^{1/2}\}b^{-1/2}\} \geq \{b^{-1/2}1b^{-1/2}\} = b^{-1}.$$

3.5.4. Proposition. *A JB algebra has an increasing approximate identity.*

Proof. Let A be a JB algebra and $\Lambda = \{u \in A : u \geq 0, \|u\| < 1\}$. Then Λ is a partially ordered set by the usual ordering on elements in A. We show Λ is directed. Let f and g be real functions defined by $f: [0, 1) \to [0, \infty)$ and $g: [0, \infty) \to [0, 1)$, and

$$f(s) = \frac{s}{1-s} = -1 + \frac{1}{1-s}, \qquad g(t) = \frac{t}{1+t} = 1 - \frac{1}{1+t}.$$

Then $f \circ g$ and $g \circ f$ are the identity functions on $[0, \infty)$ and $[0, 1)$ respectively, so that $f: \Lambda \to A_+$ and $g: A_+ \to \Lambda$ are inverses to each other. By 3.5.3 f and g are order preserving, hence Λ is order isomorphic to A_+ and therefore directed upwards.

In order to show Λ is an increasing approximate identity, for $\varepsilon>0$ let f_ε be the nonnegative real function defined on all of $\mathbb{R}$ by $f_\varepsilon(0)=0$, $f_\varepsilon(t)=1-\varepsilon$ for $|t|\geqslant\varepsilon$, and f_ε is linear on the intervals $[-\varepsilon,0]$ and $[0,\varepsilon]$. Let $a\in A$. Then $f_\varepsilon(a)$, defined via the spectral theorem (3.2.4), belongs to Λ and satisfies

$$\lim_{\varepsilon\to 0}\|a^2-\{af_\varepsilon(a)a\}\|=0. \tag{3.6}$$

Let $u\in\Lambda$, $u\geqslant f_\varepsilon(a)$. Since $u\leqslant 1$ in $\tilde{A}$, we have $a^2\geqslant\{aua\}\geqslant\{af_\varepsilon(a)a\}$, using 3.3.6. Thus by (3.6) $\lim_{\varepsilon\to 0}\|a^2-\{aua\}\|=0$; hence by 3.5.2, considering A as a subalgebra of $\tilde{A}$,

$$\begin{aligned}\|a-a\circ u\|^2&=\|a\circ(1-u)\|^2\\&\leqslant\|1-u\|\,\|\{a(1-u)a\}\|\\&=\|1-u\|\,\|a^2-\{aua\}\|\to 0.\end{aligned}$$

It follows that Λ is an increasing approximate identity for A.

3.6. States on JB algebras

3.6.1. We saw in 3.3.10 that a unital JB algebra A is an order unit space, hence the results of Section 1.2 are applicable. If A is non-unital we can still talk about positive linear functionals, namely those which map A^+ onto $\mathbb{R}^+$. If a positive functional in addition has norm 1 it is called a *state*. We shall in this section prove the Jordan algebra version of the Cauchy–Schwarz inequality together with some consequences.

3.6.2. Lemma. *Let A be a JB algebra and ρ a state on A. Then if $a,b\in A$ we have*
(i) $\rho(a\circ b)^2\leqslant\rho(a^2)\rho(b^2)$,
(ii) $\rho(a)^2\leqslant\rho(a^2)$.
In particular, the map $a\to\rho(a^2)^{1/2}$ is a seminorm on A.

Proof. Let $\lambda\in\mathbb{R}$ then we have

$$0\leqslant\rho((\lambda a+b)^2)=\lambda^2\rho(a^2)+2\lambda\rho(a\circ b)+\rho(b^2).$$

In particular, if $\rho(a^2)=0$ then $\rho(a\circ b)=0$. If $\rho(a^2)\neq 0$ let $\lambda=-\rho(a\circ b)\rho(a^2)^{-1}$, and (i) is immediate. (ii) is immediate from (i) if A is unital, letting $b=1$. Otherwise by 3.5.4 let (u_λ) be an increasing approximate identity for A. Then $\rho(a\circ u_\lambda)\to\rho(a)$ while $\rho(a^2)\rho(u_\lambda^2)\leqslant\rho(a^2)$. Thus (ii) follows from (i).

It is clear that the expression $\rho(a^2)^{1/2}$ is homogeneous in a and nonnegative. Thus in order to show it is a seminorm it remains to show the triangle inequality. However, this follows from (i) since

$$\rho((a+b)^2) \leq \rho(a^2) + 2\rho(a^2)^{1/2}\rho(b^2)^{1/2} + \rho(b^2) = (\rho(a^2)^{1/2} + \rho(b^2)^{1/2})^2.$$

3.6.3. Corollary. *Let A be a* JB *algebra and* ρ *a state on A. If* $a \in A^+$ *and* $\rho(a) = 0$ *then* $\rho(a \circ b) = 0$ *for all* $b \in A$.

Proof. If $a \geq 0$ then $0 \leq a^2 \leq \|a\|\, a$, so $\rho(a^2) = 0$. Now 3.6.2(i) implies $\rho(a \circ b) = 0$.

3.6.4. For any JB algebra A let its *state space* $S(A)$ be the set of all states on A. If A is unital then, since A is an order unit space, we know from 1.2.3 that $S(A)$ is a compact convex set. If A is not unital then compactness fails, and even the convexity of $S(A)$ is not obvious. To restore compactness, we sometimes use the set $\tilde{S}(A)$ of all positive functionals on A with norm ≤ 1 instead. Since $\tilde{S}(A)$ is characterized by $\|\rho\| \leq 1$ and $\rho(a) \geq 0$ for all $a \in A^+$, the w^* compactness of $\tilde{S}(A)$ is immediate from Alaoglu's theorem (1.1.17).

3.6.5. Lemma. $S(A)$ *is convex for any* JB *algebra* A.

Proof. Let (u_λ) be an increasing approximate identity for A (3.5.4). Let ρ be a bounded positive linear functional on A. Then $\lim \rho(u_\lambda)$ exists. For any $a \in A$ we can use the Cauchy–Schwarz inequality (3.6.2) twice to get

$$\begin{aligned} |\rho(a \circ u_\lambda)| &\leq \rho(u_\lambda^2)^{1/2}\rho(a^2)^{1/2} \leq \rho(u_\lambda^{1/2} \circ u_\lambda^{3/2})^{1/2}\,\|\rho\|^{1/2}\,\|a\| \\ &\leq \rho(u_\lambda)^{1/4}\rho(u_\lambda^3)^{1/4}\,\|\rho\|^{1/2}\,\|a\| \\ &\leq \rho(u_\lambda)^{1/4}\,\|\rho\|^{3/4}\,\|a\|. \end{aligned}$$

Taking the limits in the inequality we conclude

$$|\rho(a)| \leq [\lim \rho(u_\lambda)]^{1/4}\,\|\rho\|^{3/4}\,\|a\|,$$

which implies $\|\rho\| \leq \lim \rho(u_\lambda)$. The opposite inequality is obvious, so $\|\rho\| = \lim \rho(u_\lambda)$. From this it is clear that $S(A)$ is convex.

3.6.6. Lemma. *Let A be a* JB *algebra. Every functional* $\rho \in S(A)$ *has a unique extension* $\tilde{\rho} \in \tilde{S}(A)$. *Thus the sets* $\tilde{S}(A)$ *and* $S(\tilde{A})$ *are affinely isomorphic.*

Proof. Since any state $\tilde{\rho}$ on $\tilde{A}$ must satisfy $\tilde{\rho}(1) = 1$, uniqueness is clear. If $\rho \in \tilde{S}(A)$ define $\tilde{\rho}(a + \lambda 1) = \rho(a) + \lambda$ ($a \in A$, $\lambda \in \mathbb{R}$). Given $a \in A$ and $\lambda \in \mathbb{R}$

3.6.2(ii) implies

$$\begin{aligned}\tilde{\rho}((a+\lambda 1)^2) &= \rho(a^2)+2\lambda\rho(a)+\lambda^2\\ &\geqslant \rho(a)^2+2\lambda\rho(a)+\lambda^2=(\rho(a)+\lambda)^2\geqslant 0,\end{aligned}$$

so $\tilde{\rho}$ is positive. Since $\tilde{\rho}(1)=1$, $\tilde{\rho}$ is a state on $\tilde{A}$. Clearly, then, the restriction map $S(\tilde{A})\to\tilde{S}(A)$ is an affine isomorphism.

3.6.7. For any JB algebra A, whether unital or not, we call any extreme point of $S(A)$ a *pure state.*

3.6.8. Lemma. *Let A be a* JB *algebra. For any $a\in A$ there is a pure state ρ on A such that $|\rho(a)|=\|a\|$.*

Proof. If A is unital this follows from the corresponding result for order unit spaces (1.2.5). In general, we get a pure state $\tilde{\rho}$ on $\tilde{A}$ with $|\tilde{\rho}(a)|=\|a\|$. Let $\rho=\tilde{\rho}\mid A$. Then, by 3.6.6, ρ is an extreme point of $\tilde{S}(A)$. Then $\|\rho\|=0$ or $\|\rho\|=1$. We cannot have $\rho=0$ unless $a=0$, in which case there is nothing to prove. Thus ρ is a state, and indeed a pure one.

3.7. GNS representations of Jordan matrix algebras

3.7.1. Proposition. *Let $A=H_n(R)$, $n\geqslant 2$, be a Jordan matrix algebra which is also a* JB *algebra. Assume R is associative. Then there exists a $*$ representation of $M_n(R)$ on a complex Hilbert space carrying A isometrically onto a reversible* JC *algebra.*

Proof. Clearly $M_n(R)$ is a unital real $*$ algebra whose self-adjoint part is A, which by 3.3.10 is a complete order unit space having the identity as order unit. The proof is complete if we can show $M_n(R)$ satisfies the conditions of 1.3.10.

Let $\{e_{ij}: 1\leqslant i,j\leqslant n\}$ be the standard matrix units in $M_n(R)$. We identify each $x\in R$ with the diagonal matrix $\sum xe_{ii}$. Let $x\in R, x\neq 0$. We show $x^*x\in A^+, xx^*\in A^+$, and both are nonzero. Let $a=xe_{12}+x^*e_{21}$. Then $a\in A$, hence $0\leqslant a^2=xx^*e_{11}+x^*xe_{22}$ is nonzero. Thus xx^* and x^*x belong to A^+ and are not both zero. If, say $x^*x=0$, then $(xx^*)^2=x(x^*x)x^*=0$, so $xx^*=0$, contrary to the above. Thus both $x^*x\neq 0$ and $xx^*\neq 0$.

Let $a\in M_n(R)$. Suppose $a^*a\leqslant 0$. We show $a=0$. Let $a=\sum a_{ij}e_{ij}$, $a_{ij}\in R, i,j=1,\ldots,n$. Then $a^*=\sum a_{ji}^*e_{ij}$. If we apply the preceding paragraph successively we find, since $a^*a\leqslant 0$,

$$0\geqslant e_{ii}a^*ae_{ii}\sum_{k=1}^{n}a_{ki}^*a_{ki}e_{ii}\geqslant 0.$$

Since A is formally real (3.3.8), $a_{ki}^* a_{ki} = 0$ for all i, k, and thus $a_{ki} = 0$. In particular $a = 0$, as asserted.

Now let $a \in M_n(R)$. Then $a^* a \in A$, so by the spectral theorem (3.2.4) there are $b_1, b_2 \in A^+$ such that $b_1 \circ b_2 = 0$ and $a^* a = b_1 - b_2$. Now $(ab_2)^*(ab_2) = b_2 a^* a b_2 = -b_2^3 \leqslant 0$, so by the preceding paragraph $-b_2^3 = a^* a b_2 = 0$, hence $b_2 = 0$, and $a^* a \geqslant 0$.

It remains to show $M_n(R)$ is a Banach $*$ algebra in a suitable norm. For this define $\|a\| = \|a^* a\|^{1/2}$ for $a \in M_n(R)$. Since the norm coincides with the order norm on A we find $\|ab\|^2 = \|b^* a^* ab\| \leqslant \|b^*(\|a^* a\|)b\| = \|a\|^2 \|b\|^2$ for $a, b \in M_n(R)$. Thus $\|ab\| \leqslant \|a\| \|b\|$. If $a, b \in M_n(R)$ by 1.2.5 we have

$$\|a+b\|^2 = \|(a+b)^*(a+b)\| = \sup_{\rho \in S(A)} \rho(a^* a + a^* b + b^* a + b^* b).$$

Now each $\rho \in S(A)$ can be extended to $M_n(R)$ by letting it be zero on skew-adjoint elements. Then by the Cauchy–Schwarz inequality (1.3.10) we find

$$\begin{aligned} \|a+b\|^2 &\leqslant \sup_{\rho \in S(A)} [\rho(a^* a) + \rho(b^* b) + 2\rho(a^* a)^{1/2} \rho(b^* b)^{1/2}] \\ &\leqslant \|a\|^2 + \|b\|^2 + 2\|a\| \|b\| = (\|a\| + \|b\|)^2. \end{aligned}$$

Thus $\| \quad \|$ is a Banach algebra norm on $M_n(R)$ with respect to which $M_n(R)$ is complete since A is. We finally note that $M_n(R)$ is a Banach $*$ algebra with respect to the norm. Indeed, let $a \in M_n(R)$. Then

$$\|a\|^4 = \|a^* a\|^2 = \|a^* a a^* a\| \leqslant \|a^*(\|aa^*\|)a\| = \|aa^*\| \|a^* a\| = \|a^*\|^2 \|a\|^2.$$

Thus $\|a\| \leqslant \|a^*\|$, and by symmetry they are equal. Therefore $M_n(R)$ satisfies the conditions of 1.3.10, and the proposition follows.

3.7.2. Corollary. *Let* $A = H_n(R)$, $n \geqslant 4$, *be a Jordan matrix algebra which is also a* JB *algebra. Then there exists a* $*$ *representation of* $M_n(R)$ *on a complex Hilbert space carrying* A *isometrically onto a reversible* JC *algebra.*

Proof. By 2.7.6 R is associative, so the result follows from the proposition.

3.8. JB* algebras

3.8.1. The complex version of JB algebras are called *JB* algebras* or *Jordan C* algebras*. These algebras have applications in the theory of bounded symmetric domains. Their formal definition is as follows.

Let $\mathcal{A}$ be a complex Banach space which is a complex Jordan algebra equipped with an involution $*$. Then $\mathcal{A}$ is a *JB* algebra* if the following three conditions are satisfied for $x, y \in \mathcal{A}$:

(i) $\|x \circ y\| \leq \|x\| \|y\|$,
(ii) $\|x^*\| = \|x\|$,
(iii) $\|\{xx^*x\}\| = \|x\|^3$.

3.8.2. Proposition. *Let $\mathcal{A}$ be a* JB* *algebra. Then the set A of its self-adjoint elements is a* JB *algebra.*

Proof. If $\mathcal{A}$ is associative then $\{xx^*x\} = (x \circ x^*) \circ x$, hence $\|x\|^3 = \|\{xx^*x\}\| \leq \|x \circ x^*\| \|x\| \leq \|x\| \|x^*\| \|x\| = \|x\|^3$, so that $\|x \circ x^*\| = \|x\|^2$, and $\mathcal{A}$ is a C^* algebra. Let $\tilde{A}$ be A with the identity adjoined. Then $\tilde{A}$ is a Jordan Banach algebra (3.3.1), and for all $a \in A$ it follows by the above that $C(a)$ is isomorphic to a JB algebra. Thus as in the proof of 3.3.9 $\tilde{A}$ is an order unit space in which $-1 \leq a \leq 1$ implies $0 \leq a^2 \leq 1$. By 3.1.6 then, $\tilde{A}$ is a JB algebra in the order norm, hence so is A. Since the order norm coincides with the given norm on each $C(a)$, A is a JB algebra in the given norm.

3.8.3. It is possible to prove the converse, namely that the complexification of a JB algebra is a JB* algebra. For the much longer proof of this we refer the reader to the original paper of J. D. M. Wright [116].

3.9. Comments

JC algebras were first studied by Topping [110] and Størmer [102] (see also [41]). Their reasons were different. Topping's background was order theoretic, and he realized that much of that part of the theory of von Neumann algebras, see especially Chapter 5, could be extended to JC algebras. Størmer's studies arose from an interest in Jordan homomorphisms of C^* algebras.

The study of JB algebras was initiated by Alfsen, Shultz and Størmer [19], even though earlier approaches had been made by von Neumann [81] and Segal [88]. The axioms for JB algebras are basically streamlined versions of those given by Segal. Our treatment of JB algebras is mainly based on Alfsen *et al.* [19], except that many proofs are altered. There is, however, one major exception. In Alfsen *et al.* [19] only unital JB algebras were studied. This was remedied by Behncke [21], who proved theorem 3.3.8. The definition of JB* algebras is due to Kaplansky, who first presented it at a lecture for the Edinburgh Mathematical Society in July 1976.

In Section 3.6 we only stated the essentials about state spaces of JB algebras. Alfsen, Hanche-Olsen and Shultz [18, 15] made a detailed study of state spaces, and characterized them among all convex sets. An important ingredient in this characterization was the concept of face of a convex set. The relationship between faces of the state space and ideals of the JB algebra has been studied by Edwards [36, 37] and Hanche-Olsen [54].

We have in the present chapter defined the necessary Jordan algebraic concepts to describe the relationship between JB algebras and bounded symmetric domains. This is therefore a timely point to remark on this important aspect of JB algebras. For this we follow the survey article of Kaup [74].

Let A be a JB algebra, and let Ω denote the interior of A^+. Then it is an easy consequence of the spectral theorem that if $a \in A^+$ then $a \in \Omega$ if and only if a is invertible. In particular Ω is a non-empty open convex cone. By 3.2.10 and 3.3.6 for every $a \in \Omega$ the operator U_a is a homeomorphism of Ω which sends 1 to a^2. Therefore the U_a's, $a \in \Omega$, generate a transitive transformation group on Ω; in particular Ω is what is called a homogeneous cone. Let now $U = A \oplus \mathrm{i}A$ be the complexification of A. As remarked in 3.8.3 it follows from the work of Wright [116] that U is a JB* algebra with involution $(x+\mathrm{i}y)^* = x - \mathrm{i}y$. Let

$$D = \{z \in U: \mathrm{Im}\, z \in \Omega\},$$

where for each $z = x + \mathrm{i}y \in U$ we write $y = \mathrm{Im}\, z$. D is an example of a tube domain associated with the cone Ω, also called a generalized upper half plane for U. For example, if $A = \mathbb{R}$, $U = \mathbb{C}$, then D is the classical upper half plane. In that case the Cayley transform

$$c(z) = \mathrm{i}\frac{z-\mathrm{i}}{z+\mathrm{i}}$$

maps D biholomorphically onto the open unit disc $\Delta = \{z \in \mathbb{C}: |z| < 1\}$. In the general case we do the same, because then each $z \in D$ is invertible, and

$$c(z) = \mathrm{i}(z-\mathrm{i})(z+\mathrm{i})^{-1}$$

defines a biholomorphic map from D onto a bounded convex domain $\Delta \subset U$ called the generalized unit disc. It can be shown that Δ is the open unit ball in U.

A bounded domain B in a complex Banach space U is called symmetric if for each $a \in B$ there exists a holomorphic map $s_a: B \to B$ whose square is the identity, such that a is an isolated fixed point (s_a is uniquely determined if it exists and is called a symmetry at a). The map $s_0(z) = -z$ is the symmetry at 0 for the generalized unit disc Δ, and it can therefore be shown that Δ is symmetric. Thus D is holomorphically equivalent to a bounded symmetric domain, and is therefore a symmetric tube domain, where the latter is defined as follows. If P is an open convex cone in a real Banach space X the domain $T=\{z\in X\oplus iX: \operatorname{Im} z\in P\}$ is called a symmetric tube domain if it is biholomorphically equivalent to a bounded symmetric domain.

The converse is also true, but is much harder to prove; see Braun *et al.* [32].

Theorem. *Suppose X is a real Banach space and D is a symmetric tube domain in $U=X\oplus iX$ defined by the open convex cone Ω. Then to every $e\in\Omega$ there exists a unique Jordan product on X such that X is a* JB *algebra with unit e and generalized upper half plane D.*

Thus JB algebras, as well as JB* algebras, are in one-to-one correspondence with symmetric tube domains. The reader is referred to the notes of Loos [10] for the Jordan algebra (or Jordan triple system) characterization of more general domains.

4

JBW algebras

4.1. Topological properties

4.1.1. Recall from 1.4.5 that a C^* algebra is called a von Neumann algebra if it is also a Banach dual space. It is one of the standard results of C^* algebra theory that this is the case if and only if the algebra is monotone complete and has a separating family of normal states (for definitions see below). Furthermore, the predual is unique, and consists of all normal, bounded linear functionals. The second dual of any C^* algebra is also a von Neumann algebra; see 4.7.2 (below). The Jordan analogue of an (abstract) von Neumann algebra will be called a JBW algebra. In this chapter we shall prove the Jordan algebra analogues of the above C^* algebra results. In fact most of the results in this chapter are direct analogues of similar results for von Neumann algebras.

Let M be a JB algebra. M is said to be *monotone complete* if each bounded increasing net (a_α) in M has a least upper bound a in M. A bounded linear functional ρ on M is called *normal* if $\rho(a_\alpha) \to \rho(a)$ for each net (a_α) as above. M is said to be a *JBW algebra* if M is monotone complete and has a separating set of positive normal bounded linear functionals. We call a set of functionals *separating* if for any nonzero $a \in M$, there is a functional ρ in the set satisfying $\rho(a) \neq 0$.

We shall use the word 'state' for a positive linear functional of norm 1 even if the JB algebra is non-unital, cf. 3.6.1. In the definition of JBW algebras above we could then just as well have assumed that M has a separating set K of normal states. Since we could also have taken a larger set of normal states, we may, and shall, assume K is the set of all normal states. We denote by V the real vector subspace of M^* spanned by K.

In order to systematize notation we shall always use letters $M, N, \ldots$ for JBW algebras, and $A, B, \ldots$ for JB algebras.

4.1.2. An alternative description of von Neumann algebras is that they are ultraweakly closed C^* algebras on a Hilbert space; cf. 1.4.6. The corresponding Jordan algebras will be called JW algebras. This will turn out to be a smaller class than that of JBW algebras, as there are exceptional JBW algebras like $H_3(\mathbb{O})$; see 2.8.5 and 3.1.7.

Let H be a complex Hilbert space and $B(H)$ the von Neumann algebra of all bounded operators on H equipped with the ultraweak topology; see 1.4. A Jordan subalgebra M of $B(H)_{sa}$ is called a *JW algebra* if M is ultraweakly closed. More generally, we shall sometimes use the term JW algebra to mean a JB algebra which is isomorphic to a JW algebra like M above. It will be a consequence of 1.4.4 and 4.4.16 below that a JW algebra as defined above is indeed a JBW algebra. This is, however, a trivial observation for the reader who knows that the ultraweakly continuous linear functionals on $B(H)$ are normal, and that the least upper bound of a bounded increasing net in $B(H)_{sa}$ is the ultraweak limit of the net.

4.1.3. Let M be a JBW algebra with set of normal states denoted by K and V the vector space spanned by K. Then the *weak topology* on M is the $\sigma(M, V)$ topology; see (1.1.7). The *strong topology* on M is the locally convex topology defined by the seminorms $a \to \rho(a^2)^{1/2}$, $\rho \in K$; see 3.6.2.

If $\rho \in K$ then $\rho(a^2)^{1/2} \leqslant \|a\|$, hence norm convergence implies strong convergence. From the Cauchy–Schwarz inequality 3.6.2(ii) $\rho(a^2)^{1/2} \to 0$ implies $\rho(a) \to 0$, hence strong convergence implies weak convergence. Furthermore, if (a_α) is a bounded increasing net in M with least upper bound a in M, written $a_\alpha \nearrow a$, then (a_α) converges strongly to a. Indeed, if $\rho \in K$ then $a - a_\alpha \geqslant 0$, so that

$$\rho((a-a_\alpha)^2) \leqslant \|a-a_\alpha\|\, \rho(a-a_\alpha) \leqslant \|a\|\, \rho(a-a_\alpha) \to 0.$$

In the second inequality we assumed $a_\alpha \geqslant 0$, as we may without loss of generality.

4.1.4. Recall that if $a \in M$ then T_a is the operator $b \to a \circ b$ and U_a the operator $b \to \{aba\}$. By a straightforward computation we find

$$T_a = \tfrac{1}{2}(U_{1+a} - U_a + \iota),$$

where 1 is the identity in $\tilde{M}$ and ι the identity map. We denote by T_a^* and U_a^* the adjoint maps on M^*; cf. 1.1.20.

4.1.5. Lemma. *Let M be a JBW algebra with normal states K and vector space V spanned by K. Then for all $a \in M$, $T_a^*(V) \subset V$ and $U_a^*(V) \subset V$.*

Proof. By 4.1.4 it suffices to show $U_a^*(K) \subset V^+$, where V^+ denotes the positive linear functionals in V. Suppose $b_\alpha \nearrow b$ in M, and let $\rho \in K$. Then $(U_a^*\rho)(c) = \rho(U_a c)$ for each $c \in M$, so it suffices to show $U_a b_\alpha \nearrow U_a b$. If a

is invertible, by 3.2.10 and 3.3.6 U_a and $U_a^{-1} = U_{a^{-1}}$ are positive maps, so U_a is an order automorphism of M, and the result follows. For arbitrary a choose $\lambda > 0$ in $\mathbb{R}$ such that both $\lambda 1 + a$ and $\lambda 1 - a$ are invertible in $\tilde{M}$. Then by the above we have

$$U_{\lambda 1 \pm a} b_\alpha \nearrow U_{\lambda 1 \pm a} b. \tag{4.1}$$

It follows from the definition of $U_{\lambda 1 \pm a}$ that $U_{\lambda 1 \pm a} = \lambda^2 \iota \pm 2\lambda T_a + U_a$, hence $U_{\lambda 1 + a} + U_{\lambda 1 - a} = 2\lambda^2 \iota + 2U_a$. Thus by (4.1) $2\lambda^2 b_\alpha + 2U_a b_\alpha \nearrow 2\lambda^2 b + 2U_a b$. Since by assumption $b_\alpha \nearrow b$ and $U_a b_\alpha \leqslant U_a b$ we obtain $U_a b_\alpha \nearrow U_a b$.

4.1.6. Corollary. *Let M be a* JBW *algebra. Then T_a and U_a are weakly continuous for all $a \in M$.*

4.1.7. Lemma. *A* JBW *algebra M is unital.*

Proof. Let by 3.5.4 (u_λ) be an increasing approximate identity for M. Then (u_λ) is a bounded increasing net, so has a least upper bound e in M. By 4.1.3 $u_\lambda \to e$ strongly, hence weakly. If $a \in M$ then by 4.1.6 $u_\lambda \circ a \to e \circ a$ weakly. But $u_\lambda \circ a \to a$ in norm, hence weakly by 4.1.3. Thus $e \circ a = a$, and e is the identity in M.

4.1.8. Lemma. *Let M be a* JBW *algebra. Then T_a and U_a are strongly continuous for all $a \in M$.*

Proof. Let by 4.1.7 1 denote the identity in M. By 4.1.4 it suffices to show that U_a is strongly continuous. Suppose $b_\alpha \to 0$ strongly. Then $\rho(b_\alpha^2) \to 0$ for all $\rho \in K$. Since by 4.1.5 and the positivity of U_a (3.3.6) $\rho(a^2)^{-1} \rho \circ U_a \in K$ whenever $\rho(a^2) \neq 0$, and $\rho \circ U_a = 0$ if $\rho(a^2) = 0$, we have $\rho(U_a b_\alpha^2) \to 0$. But by 2.4.17 and the positivity of U_a, if $b \in M$ then

$$(U_a b)^2 = \{a\{ba^2b\}a\} \leqslant \|a\|^2 \{a\{b1b\}a\} = \|a\|^2 U_a b^2.$$

Thus $\rho((U_a b_\alpha)^2) \to 0$, and $U_a b_\alpha \to 0$ strongly.

4.1.9. Lemma. *Multiplication is jointly strongly continuous on bounded subsets of a* JBW *algebra.*

Proof. Let (a_α) and (b_β) be bounded nets in a JBW algebra M. Suppose first $a_\alpha \to 0$ strongly. From the inequalities $0 \leqslant (a_\alpha^2)^2 \leqslant \|a_\alpha^2\| a_\alpha^2$ we find that $a_\alpha^2 \to 0$ strongly. If furthermore $b_\beta \to 0$ strongly then from the identity $a \circ b = \frac{1}{2}((a+b)^2 - a^2 - b^2)$ and the fact that addition is strongly continuous by definition of the strong topology, it follows that $a_\alpha \circ b_\beta \to 0$ strongly.

Let now $a_\alpha \to a$ and $b_\beta \to b$ strongly. Then we have the identity $a \circ b - a_\alpha \circ b_\beta = (a - a_\alpha) \circ b + (a_\alpha - a) \circ (b - b_\beta) + a \circ (b - b_\beta)$. From the previous paragraph the middle term converges strongly to 0, while by 4.1.8 the other two terms converge strongly to 0.

4.1.10. Let M be a JBW algebra and $a \in M$. We denote by $W(a)$ the weak closure of $C(a)$, where the latter is defined in 3.2.3.

If $b, c \in W(a)$ choose nets (b_α) and (c_β) in $C(a)$ which converge weakly to b and c respectively. By 4.1.6 we get $b_\alpha \circ c_\beta \to b \circ c_\beta$ weakly with α, so $b \circ c_\beta \in W(a)$. Again we find $b \circ c_\beta \to b \circ c$ weakly, so $b \circ c \in W(a)$. Thus $W(a)$ is a subalgebra of M. Note that since the norm topology is finer than the weak topology, $W(a)$ is norm closed, hence is a JB algebra. If (b_α) is an increasing net in $W(a)$ with least upper bound b then $b_\alpha \to b$ strongly, hence weakly. Thus $b \in W(a)$, and $W(a)$ is monotone complete. It is thus a JBW algebra, since the states in K restrict to normal states on $W(a)$. Finally we note that a modification of the approximation argument above shows that the associativity of $C(a)$ implies that $W(a)$ is associative. By an application of 3.2.2 we can conclude:

4.1.11. Lemma. *Let M be a* JBW *algebra and $a \in M$. Then $W(a)$ is an associative* JBW *subalgebra of M isometrically isomorphic to a monotone complete $C(X)$, where X is a compact Hausdorff space.*

4.1.12. An idempotent in a JB algebra will be called a *projection*. Just as for von Neumann algebras projections will play an important role in the theory of JBW algebras.

4.1.13. Lemma. *Let M be a* JBW *algebra and p a nonzero projection in M. Let K denote the set of normal states on M. Then $M_p = U_p(M)$ is a* JBW *subalgebra of M with set of normal states $K_p = \{\rho \mid M_p ; \rho \in K, \rho(p) = 1\}$. Furthermore, if $a \in M_p$, $b \in M$ and $0 \leq b \leq a$ then $b \in M_p$.*

Proof. By 2.5.5 M_p is a subalgebra of M. It is weakly closed, for if (a_α) is a net in M and $U_p a_\alpha \to a$ weakly, then by weak continuity of U_p (4.1.6) together with the formula $U_p = U_{p^2} = U_p^2$ (2.5.4), we have

$$U_p a = \text{weak} \lim_\alpha U_p(U_p a_\alpha)) = \text{weak} \lim_\alpha U_p a_\alpha = a,$$

so $a \in M_p$. By 3.3.6 and 4.1.8 U_p is positive and strongly continuous. Thus if $a_\alpha \nearrow a$ in M then $U_p a_\alpha \nearrow U_p a$, so if $a_\alpha \in M_p$ then $a \in M_p$, hence M_p is a monotone complete JB algebra.

If $\rho \in K$ and $\rho(p) \neq 0$ then clearly $\rho(p)^{-1}\rho \mid M_p$ is a normal state on M_p. In particular K_p is a separating set of normal states on M_p, so that M_p is a JBW algebra. Let ω be a normal state on M_p; then there is $\rho \in K$ such that $\omega = \rho \mid M_p$. Indeed, let $\rho = U_p^*\omega$. Since U_p is positive and preserves monotone limits ρ is normal, hence $\rho \in K$, and we have shown K_p is as asserted.

Finally suppose $0 \leq b \leq a$ with $a \in M_p$, $b \in M$. By 3.3.6 $0 \leq U_{p^\perp} b \leq U_{p^\perp} a = 0$, so that $U_{p^\perp} b = 0$. Since $0 \leq b^2 \leq \|b\|\, b$ we have in particular $0 \leq U_{p^\perp} b^2 \leq \|b\|\, U_{p^\perp} b = 0$, and so $U_{p^\perp} b^2 = 0$. From 2.4.17 $(U_x y)^2 = U_x U_y x^2$ for all x, y, hence we have $U_b p^\perp = 0$. By Macdonald's theorem (2.4.13) we also have the following identity:

$$4(x \circ y)^2 = 2x \circ U_y x + U_x y^2 + U_y x^2.$$

Thus $(p^\perp \circ b)^2 = 0$, hence $p \circ b = b$. But then $U_p b = 2p \circ (p \circ b) - p \circ b = b$, and $b \in M_p$.

4.1.14. For later reference we note that in the final paragraph of the preceding proof we actually proved that if p is a projection and a a positive element in a JB algebra, then $\{pap\} = 0$ implies $p \circ a = 0$.

4.2. Projections in JBW algebras

4.2.1. Just as in the previous section we denote by $W(a)$ the JBW algebra generated by an element a and 1 in a JBW algebra M. By 4.1.11 it is quite straightforward to modify existing theory for von Neumann algebras to obtain the usual integral form of the spectral theorem for $a \in M$. We shall not need that much and will satisfy ourselves with a weaker result. Recall that two projections p and q are said to be orthogonal, written $p \perp q$, if $p \circ q = 0$.

4.2.2. Lemma. *Let A be a JB algebra and p, q projections in A. Then the following five conditions are equivalent:*

(i) *p and q are orthogonal.*
(ii) *$p + q$ is a projection.*
(iii) $p + q \leq 1$.
(iv) $U_p q = 0$.
(v) $U_p U_q = 0$.

Proof. (i)$\Rightarrow$(ii) follows since $p \circ q = 0$ implies $(p+q)^2 = p^2 + q^2 = p + q$.

(ii)$\Rightarrow$(iii) is immediate since each projection is majorized by 1.

(iii)$\Rightarrow$(iv) follows by positivity of U_p (3.3.6) since $p \leq p + U_p q = U_p p + U_p q = U_p(p+q) \leq p$, so that $U_p q = 0$.

(iv)$\Rightarrow$(v) If $a \in M_+$ then $0 \leq U_q a \leq \|a\|\, q$, hence by (iv) $0 \leq U_p U_q a \leq \|a\|\, U_p q = 0$, so $U_p U_q a = 0$. Since M is linearly spanned by its positive elements, (v) follows.

(v)$\Rightarrow$(i) follows from the fact that $U_p q = U_p U_q 1 = 0$, and so $p \circ q = 0$ by 4.1.14.

4.2.3. Proposition. *Let M be a* JBW *algebra, $a \in M$, and $\varepsilon > 0$. Then there exist pairwise orthogonal projections $p_1, \ldots, p_n$ in $W(a)$ and real numbers $\lambda_1, \ldots, \lambda_n$ such that*

$$\left\| a - \sum_{i=1}^{n} \lambda_i p_i \right\| < \varepsilon.$$

Proof. By 4.1.11 $W(a)$ is isometrically isomorphic to $C(X)$ with $C(X)$ monotone complete. We shall therefore identify $W(a)$ with $C(X)$ and write elements b in $W(a)$ as functions $b(x)$. It follows from the monotone completeness of $C(X)$ that X is Stonean, i.e. the closure of an open set is open. However, we shall not use this. Let $\lambda \in \mathbb{R}$ and let Ω be the set of $x \in X$ where $a(x) < \lambda$. We shall prove directly that $\bar{\Omega}$ is open. Let $f_n : \mathbb{R} \to \mathbb{R}$ be the continuous function such that $f_n(\mu) = 1$ when $\mu \leq \lambda - 1/n$, $f_n(\mu) = 0$ when $\mu \geq \lambda$, and f_n is linear in $[\lambda - 1/n, \lambda]$. Then $(f_n(a))$ is an increasing bounded net in $C(X)$, and hence has a least upper bound $e \in C(X)$. Clearly, $0 \leq e \leq 1$. If $x \in \Omega$ then for large n, $f_n(a(x)) = 1$, hence $1 \leq e(x) \leq 1$. By the continuity of e, $e(x) = 1$ for all $x \in \bar{\Omega}$. If $x \notin \bar{\Omega}$ there is some $b \in C(X)$ such that $0 \leq b \leq 1$, $b(x) = 0$, $b|_{\bar{\Omega}} = 1$. Then $b \geq f_n(a)$ for all n, so $b \geq e$, and $0 \leq e(x) \leq b(x) = 0$. Thus the continuous function e is the characteristic function of $\bar{\Omega}$, which is therefore open and closed.

Pick now $\lambda_0 < \lambda_1 < \ldots < \lambda_n$ such that $\lambda_0 \leq a \leq \lambda_n$ and $\lambda_{i+1} - \lambda_i < \varepsilon$ for $i = 0, \ldots, n-1$. By the above, there are open and closed subsets E_i such that $a \leq \lambda_i$ on E_i while $a \geq \lambda_i$ outside E_i. Clearly, $\emptyset = E_0 \subseteq E_1 \subseteq \ldots \subseteq E_n = X$. Let p_i be the characteristic function of $E_i \setminus E_{i-1}$, $i = 1, \ldots, n$. Then, if $b = \sum_{i=1}^{n} \lambda_i p_i$, $b|_{E_i \setminus E_{i-1}} = \lambda_i$, and it easily follows that $\|a - b\| < \varepsilon$.

4.2.4. Recall from 2.5.1 that two elements a and b in a Jordan algebra A operator commute if $T_a T_b = T_b T_a$. We showed in 2.5.5 that if A is unital and p is an idempotent in A then a and p operator commute if and only if p and a generate an associative algebra, if and only if $U_p a = p \circ a$. We next extend this result for JBW algebras.

4.2.5. Lemma. *Let M be a* JBW *algebra and $a \in M$. If p is a projection in M the following conditions are equivalent.*

(i) *p operator commutes with a.*
(ii) *p operator commutes with all elements in $W(a)$.*
(iii) *p operator commutes with all projections in $W(a)$.*

Proof. (i)$\Rightarrow$(ii) If p operator commutes with a then p and a generate an associative Jordan algebra (2.5.5), hence by the argument of 4.1.10 the weakly closed algebra generated by p and a is associative. Since this algebra contains $W(a)$, again by 2.5.5 p operator commutes with all elements of $W(a)$.

(ii)$\Rightarrow$(iii) is trivial.

(iii)$\Rightarrow$(i) If p operator commutes with all projections in $W(a)$ it operator commutes with all their linear combinations. By norm continuity of T_p and U_p (3.2.7) p operator commutes with a by 4.2.3.

4.2.6. Lemma. *Let M be a JBW algebra and $a \in M$. Then there exists a smallest projection $r(a)$ in M with the property that $r(a) \circ a = a$. $r(a)$ belongs to $W(a)$.*

We call $r(a)$ the *range projection* of a.

Proof. For each $n \in \mathbb{N}$ let f_n be the real continuous function defined by $f_n(t) = 1$ for $|t| \geq 1/n$, $f_n(0) = 0$, and f_n is linear on the intervals $[-1/n, 0]$ and $[0, 1/n]$. Then (f_n) is an increasing sequence of functions bounded by 1. As in the proof of 4.2.3 the sequence $(f_n(a))$ is increasing and converges strongly to its least upper bound $r(a)$. Using 4.1.9 it is easy to see that $(f_n(a)^2)$ converges strongly to $r(a)$ as well, hence $r(a)$ is a projection in $W(a)$. Since $\lim_n \|f_n(a) \circ a - a\| = 0$, $r(a) \circ a = a$.

Suppose p is a projection in M such that $p \circ a = a$. Then $U_p a = 2p \circ (p \circ a) - p \circ a = p \circ a$, so that p and a operator commute by 2.5.5. By 4.2.5 p, and hence $1-p$, operator commutes with all elements in $W(a)$. If J denotes the set of $b \in W(a)$ such that $(1-p) \circ b = 0$ then by 2.5.8 J is an ideal in $W(a)$. Now, by the Stone–Weierstrass theorem, polynomials without constant terms are dense among the functions in $C(\mathrm{Sp}\, a)$ vanishing at 0. Since $a \in J$ it therefore follows from the spectral theorem (3.2.4) that J contains all elements $f(a)$ in $C(a)$, where f is a continuous real function on $\mathbb{R}$ such that $f(0) = 0$. In particular $f_n(a) \in J$ for all n. Since J is strongly closed by 4.1.8, $r(a) \in J$. Thus $r(a) = p \circ r(a)$, so that $r(a) = \{p r(a) p\} \leq p$, proving the minimality of $r(a)$.

4.2.7. If M is a JBW algebra we use the symbols $\bigvee$ and $\bigwedge$ to denote the least upper bound and greatest lower bound in the set of projections in M if they exist, as we now proceed to show that they do.

4.2.8. Lemma. *Let M be a JBW algebra and P the set of projections in M.*

(i) *If $p_1, \ldots, p_n \in P$ then* $\bigvee_{i=1}^{n} p_i = r\left(\sum_{i=1}^{n} p_i\right)$ *and* $\bigwedge_{i=1}^{n} p_i = \left(\bigvee_{i=1}^{n} p_i^{\perp}\right)^{\perp}$.

(ii) *If $(p_\alpha)_{\alpha \in I}$ is a family of projections in P then $\bigvee_{\alpha \in I} p_\alpha$ and $\bigwedge_{\alpha \in I} p_\alpha$ exist.*

Proof. (i) By 4.1.13 $0 \leqslant a \leqslant b$ implies $r(a) \leqslant r(b)$. Therefore

$$p_j = r(p_j) \leqslant r\left(\sum_{i=1}^{n} p_i\right),$$

so that

$$\bigvee_{i=1}^{n} p_i \leqslant r\left(\sum_{i=1}^{n} p_i\right).$$

If $q \in P$ and $p_i \leqslant q$, $i = 1, \ldots, n$, then $\sum_{i=1}^{n} p_i \leqslant nq$, so that $r(\sum_{i=1}^{n} p_i) \leqslant q$. This proves

$$\bigvee_{i=1}^{n} p_i = r\left(\sum_{i=1}^{n} p_i\right).$$

Let $q_i = p_i^{\perp}$. By the above

$$\left(\bigvee_{i=1}^{n} q_i\right)^{\perp} = 1 - r\left(\sum_{i=1}^{n} q_i\right) \leqslant 1 - q_j = p_j$$

for all j, hence

$$\left(\bigvee_{i=1}^{n} q_i\right)^{\perp} \leqslant \bigwedge_{i=1}^{n} p_i.$$

Conversely, since $\bigwedge_{i=1}^{n} p_i \leqslant p_j$ we have

$$q_j = p_j^{\perp} \leqslant \left(\bigwedge_{i=1}^{n} p_i\right)^{\perp},$$

hence

$$\left(\bigvee_{i=1}^{n} q_i\right) \leqslant \left(\bigwedge_{i=1}^{n} p_i\right)^{\perp},$$

so that

$$\bigwedge_{i=1}^{n} p_i \leqslant \left(\bigvee_{i=1}^{n} q_i\right)^{\perp},$$

and they are equal.

(ii) If $(p_\alpha)_{\alpha \in I}$ is a subfamily of P let $\mathscr{F}$ be the directed set of finite subsets of I ordered by inclusion. For $F \in \mathscr{F}$ let $p_F = \bigvee_{i \in F} p_i$. By (i) $p_F \in P$, and $(p_F)_{F \in \mathscr{F}}$ is an increasing net in P, which converges strongly to its least upper bound p. By strong continuity of multiplication on bounded sets in M (4.1.9) p is a projection, so lies in P. Thus $\bigvee_{\alpha \in I} p_\alpha$ exists in P.

Finally the same argument as for $\bigvee_{\alpha \in I} p_\alpha$ in the previous paragraph applies to $\bigwedge_{\alpha \in I} p_\alpha$ and shows that $\bigwedge_{\alpha \in I} p_\alpha \in P$.

4.2.9. Remark. Let M be a JBW algebra and $(a_\alpha)_{\alpha\in I}$ an indexed family of elements in M with $\|a_\alpha\|\leq k<\infty$ for all $\alpha\in I$ and with pairwise orthogonal range projections $r(a_\alpha)$. Then the infinite sum $\sum_{\alpha\in I} a_\alpha$ exists in M as the strong limit of the bounded net $(\sum_{\alpha\in F} a_\alpha)_{F\in\mathcal{F}}$, where $\mathcal{F}$ is the set of finite subsets of I ordered by inclusion. Moreover, $\|\sum_{\alpha\in I} a_\alpha\|\leq k$. In particular, if $(p_\alpha)_{\alpha\in I}$ is a family of pairwise orthogonal projections in M then $\sum_{\alpha\in I} p_\alpha$ is a projection in M, indeed, $\sum_{\alpha\in I} p_\alpha = \bigvee_{\alpha\in I} p_\alpha$.

To prove the above note that by the spectral theorem (3.2.4) each $a_\alpha = a_\alpha^+ - a_\alpha^-$ with $\|a_\alpha^\pm\|\leq k$ and $a_\alpha^+\circ a_\alpha^- = 0$. By strong continuity of addition we may consider the two families $(a_\alpha^+)_{\alpha\in I}$ and $(a_\alpha^-)_{\alpha\in I}$ separately, hence we may assume $0\leq a_\alpha\leq k1$ for all $\alpha\in I$. Since the range projections $r(a_\alpha)$ are pairwise orthogonal, it follows from 4.2.2 that if $F\in\mathcal{F}$ then $0\leq\sum_{\alpha\in F} a_\alpha\leq\sum_{\alpha\in F} kr(a_\alpha)\leq k1$. Then the net $(\sum_{\alpha\in F} a_\alpha)_{F\in\mathcal{F}}$ is bounded and increasing, so by 4.1.3 converges strongly to its least upper bound $\sum_{\alpha\in I} a_\alpha$. Note that since multiplication is strongly continuous on bounded sets (4.1.9), if $a_\alpha = p_\alpha$ are pairwise orthogonal projections then by 4.2.2 each sum $\sum_{\alpha\in F} a_\alpha$ is a projection, hence so is their strong limit $\sum_{\alpha\in F} p_\alpha$. By 4.2.8 it is clear that $\sum_{\alpha\in I} p_\alpha = \bigvee_{\alpha\in I} p_\alpha$.

4.3. The centre

4.3.1. By 2.5.1 the centre of a Jordan algebra A is the set of elements in A which operator commute with all elements in A. Since by 4.1.6 multiplication is separately weakly continuous in a JBW algebra M, the centre Z of M is an associative JBW subalgebra of M. We shall often write ab instead of $a\circ b$ if $a\in Z$ and $b\in M$.

Recall that a symmetry in M is an operator s such that $s^2=1$. The following characterization of the centre is quite useful.

4.3.2. Lemma. *Let M be a* JBW *algebra with centre Z. If $a\in M$ then $a\in Z$ if and only if $U_s a = a$ for all symmetries $s\in M$.*

Proof. There is a one-to-one correspondence between the set of projections in M and the set of symmetries in M given by $p\to s=2p-1$. Furthermore it is easily verified that $U_{2p-1}=2U_p+2U_{1-p}-\iota$. By 4.2.5 $a\in Z$ if and only if a operator commutes with all projections in M, hence by 2.5.5 if and only if $a=U_p a+U_{1-p}a$ for all projections p in M. By the above identity this is equivalent to $U_s a = a$ for all symmetries $s\in M$.

4.3.3. Let M be a JBW algebra with centre Z. If p is a projection in M then its *central support* $c(p)$ is the smallest projection in Z majorizing p.

4.3.4. Proposition. *Let M be a JBW algebra with centre Z. Let p be a projection in M. Then the map $T_p: Z \to Zp$ is a homomorphism with kernel $Z(1-c(p))$.*

Proof. The map T_p is by 2.5.8 and 4.1.6 a weakly continuous homomorphism of the associative JBW algebra Z onto Zp. In particular its kernel is a weakly closed ideal J in Z. By 4.2.3 J is generated by its projections. If e is a projection in J then $(1-e)p=p$, so $1-e \geq c(p)$, hence $e \leq 1-c(p)$. Since conversely $(1-c(p))p=0$, the proof is complete.

4.3.5. Lemma. *Let M be a JBW algebra with centre Z. If p is a projection in M and e a projection in Z then $c(ep)=ec(p)$.*

Proof. Clearly $c(ep) \leq ec(p)$. Conversely since $((1-c(ep))e)p = (1-c(ep))ep=0$ it follows from 4.3.4 that $(1-c(ep))ec(p)=0$. But then $ec(p) \leq c(ep)$, and they are equal.

4.3.6. Proposition. *Let M be a JBW algebra and J a weakly closed ideal in M. Then there exists a central projection e in M such that $J=eM$.*

Proof. Since J is itself a JBW algebra, it has by 4.1.7 an identity e. Thus if $a \in M$ then $e \circ a \in J$, and so $e \circ a = e \circ (e \circ a)$, so that e is central in M by 2.5.5, proving the proposition.

4.3.7. Remark. If ϕ is a weakly continuous homomorphism of a JBW algebra M into another JBW algebra, then its kernel is a weakly closed ideal, hence of the form eM. The projection $e^{\perp}$ is called the *support* of ϕ, and is the smallest projection f in M such that $\phi(f)=\phi(1)$.

4.3.8. Lemma. *Let $\mathcal{M}$ be a von Neumann algebra and M a weakly closed Jordan subalgebra of $\mathcal{M}_{sa}$, and assume that M generates $\mathcal{M}$. Then the centre of M is contained in the centre of $\mathcal{M}$.*

Proof. By 4.2.3, since the centre of M is a JBW algebra it is enough to show that any central projection e in M belongs to the centre of $\mathcal{M}$. If $a \in M$ then by 2.5.5 $a = eae + e^{\perp}ae^{\perp}$, from which $ea = ae$ follows. Since M generates $\mathcal{M}$, $ea = ae$ for all $a \in \mathcal{M}$, so e is in the centre of $\mathcal{M}$.

4.4. The second dual

4.4.1. A useful result on C^* algebras is the observation that the second dual of a C^* algebra is a von Neumann algebra. Furthermore, a C^*

algebra is a von Neumann algebra if and only if it is a dual space. We shall in the present section prove the analogous results for JB algebras. In the process we shall make heavy use of the results and concepts of Sections 1.1 and 1.2. Our first result is a weak form of the GNS representation for states on C^* algebras.

4.4.2. Lemma. *Let A be a* JB *algebra and ρ a state on A. Then there exists a real Hilbert space H_ρ with inner product $\langle\ ,\ \rangle$ and a linear map $\eta_\rho : A \to H_\rho$ with $\|\eta_\rho\| \leq 1$ such that $\eta_\rho(A)$ is dense in H_ρ and $\rho(a \circ b) = \langle \eta_\rho(a), \eta_\rho(b)\rangle$ for all $a, b \in A$.*

Proof. Define a bilinear form on A by $(a, b) = \rho(a \circ b)$. Let $N = \{a \in A : (a, a) = 0\}$. By the Cauchy–Schwarz inequality (3.6.3) $a \in N$ if and only if $(a, b) = 0$ for all $b \in A$. Thus N is a linear subspace of A, and A/N is a real pre-Hilbert space with completion H_ρ. If η_ρ denotes the quotient map $\eta_\rho : A \to A/N \subset H_\rho$ then $\rho(a \circ b) = (a, b) = \langle \eta_\rho(a), \eta_\rho(b)\rangle$ for all $a, b \in A$, and $\|\eta_\rho(a)\|^2 = \rho(a^2) \leq \|a^2\| = \|a\|^2$, so $\|\eta_\rho\| \leq 1$.

4.4.3. Theorem. *Let A be a* JB *algebra. Then its second dual A^{**} is a* JBW *algebra whose product extends the original product in A and which is separately weak-$*$ continuous. Furthermore, the weak-$*$ continuous extension to A^{**} of a state on A is normal.*

Proof. We first assume A is unital. Let $S(A)$ be the state space of A (1.2.1) and let $\rho \in S(A)$. By 4.4.2 there is a real Hilbert space H_ρ and a linear map $\eta_\rho : A \to H_\rho$ with $\|\eta_\rho\| \leq 1$ such that $\eta_\rho(A)$ is dense in H_ρ and

$$\rho(a \circ b) = \langle \eta_\rho(a), \eta_\rho(b)\rangle, \qquad a, b \in A. \tag{4.2}$$

Since Hilbert spaces can be identified with their duals, $H_\rho = H_\rho^{**}$. By 1.1.21 let η_ρ^{**} denote the extension of η_ρ mapping A^{**} into $H_\rho^{**} = H_\rho$. Then $\|\eta_\rho^{**}\| \leq 1$. If $a, b \in A^{**}$ define $f_{a,b}: S(A) \to \mathbb{R}$ by

$$f_{a,b}(\rho) = \langle \eta_\rho^{**}(a), \eta_\rho^{**}(b)\rangle. \tag{4.3}$$

Since by 1.1.21 η_ρ^{**} is continuous when A^{**} is given the weak-$*$ topology and H_ρ the weak topology, $f_{a,b}(\rho)$ is separately weak-$*$ continuous in the variables a, b in A^{**} for each $\rho \in S(A)$. By (4.2) and (4.3) if $a,b \in A$, $\rho \in S(A)$ then $f_{a,b}(\rho) = \rho(a \circ b)$. Thus $f_{a,b}$ is a w^*-continuous affine function on $S(A)$ when $a,b \in A$. Now A is weak-$*$ dense in A^{**} (1.1.19). Hence if $a, b \in A^{**}$ we can choose nets (a_α) and (b_β) in A converging weak-$*$ to a and b respectively. Then we have

$$f_{a,b}(\rho) = \lim_\alpha f_{a_\alpha, b}(\rho) = \lim_\alpha \lim_\beta f_{a_\alpha, b_\beta}(\rho),$$

so that $f_{a,b}$ is an affine function on $S(A)$. It is also bounded since

$|f_{a,b}(\rho)| \leqslant \|\eta_\rho^{**}(a)\| \, \|\eta_\rho^{**}(b)\| \leqslant \|a\| \, \|b\|$. By 1.2.11 $f_{a,b}$ can thus be extended to a unique element $a \circ b$ in A^{**} satisfying $(a \circ b)(\rho) = f_{a,b}(\rho)$, $\rho \in S(A)$. Then the map $A^{**} \times A^{**} \to A^{**}$ defined by $(a, b) \to a \circ b$ is separately weak-$*$ continuous in a and b.

Recall from 2.4.3 that the linearized Jordan axiom is

$$([T_{a \circ b}, T_c] + [T_{b \circ c}, T_a] + [T_{c \circ a}, T_b])(d) = 0$$

for $a, b, c, d \in A$. If we take weak-$*$ limits in each variable a, b, c, d separately we see that the identity also holds in A^{**}. Since commutativity is obvious, A^{**} is a Jordan algebra.

By 3.3.10 A is a complete order unit space such that $-1 \leqslant a \leqslant 1$ implies $0 \leqslant a^2 \leqslant 1$. It follows from 1.2.7 that A^{**} is a complete order unit space such that if $a \in A^{**}$ and $-1 \leqslant a \leqslant 1$ then $\|a\| \leqslant 1$. But then, if $\rho \in S(A)$, we have

$$a^2(\rho) = f_{a,a}(\rho) = \|\eta_\rho^{**}(a)\|^2 \leqslant \|a\|^2 \leqslant 1,$$

so that $0 \leqslant a^2 \leqslant 1$. Thus A^{**} is a JB algebra by 3.1.6. Furthermore, A^{**} is monotone complete since it is by 1.2.11 order isomorphic to $A^b(S(A))$.

In order to show that A^{**} is a JBW algebra it remains to exhibit a separating set of normal states. Let $\rho \in S(A)$ and let $\bar{\rho}$ be its weak-$*$ continuous extension to A^{**} (1.1.22). Suppose (a_α) is a bounded increasing net in A^{**} with least upper bound a in A^{**}. Since A^{**} is order isomorphic to $A^b(S(A))$, and since $a_\alpha \nearrow a$, $a_\alpha \mid S(A) \to a \mid S(A)$ pointwise. In particular $\bar{\rho}(a_\alpha) = a_\alpha(\rho) \to a(\rho) = \bar{\rho}(a)$, so that $\bar{\rho}$ is a normal state on A^{**}. Since $S(A)$ separate points on A^{**} we have shown that A^{**} is a JBW algebra whenever A is unital.

Finally, if A is non-unital we can apply the previous results to $\tilde{A}$. Then by 1.1.23 and the previous discussion A^{**} is a weak-$*$ closed subalgebra of $(\tilde{A})^{**}$, hence A^{**} is in particular a JB algebra. It is monotone complete, since if (a_α) is a bounded increasing net in A^{**} with least upper bound a in $(\tilde{A})^{**}$ then for all $\rho \in S(\tilde{A})$, $\bar{\rho}(a_\alpha) \to \bar{\rho}(a)$. Thus $a_\alpha \to a$ weak-$*$, proving that $a \in A^{**}$. Any state ρ_0 of A extends to a state ρ of $\tilde{A}$ by $\rho(1) = 1$ (3.6.6). Then $\bar{\rho}$ restricts to a weak-$*$ continuous state $\bar{\rho}_0$ on A^{**} extending ρ. Since we have seen that any bounded increasing net in A^{**} has a least upper bound in A^{**} coinciding with its least upper bound in $(\tilde{A})^{**}$, $\bar{\rho}_0$ is normal. Thus A^{**} has a separating set of normal states, so it is a JBW algebra.

4.4.4. From 4.4.3 it immediately follows that weak convergence in A^{**}, as defined in 4.1.3, implies weak-$*$ convergence. For, by 1.1.8 every weak-$*$ continuous linear functional on A^{**} is $\bar{\rho}$ for some $\rho \in A^*$. Since by 1.2.6 ρ is a linear combination $\rho = \lambda\sigma + \mu\tau$ of states, it follows from 4.4.3 that $\bar{\rho} = \lambda\bar{\sigma} + \mu\bar{\tau}$ is a linear combination of normal states on A^{**}.

It will follow from 4.4.16 below that the weak and weak-$*$ topologies coincide on A^{**}.

For the moment, however, we shall note that any weak-$*$ closed ideal in A^{**} is weakly closed, hence by 4.3.6 is of the form eA^{**} for a central projection in A^{**}.

4.4.5. Our next objective is to show A^{**} is the monotone completion of A, i.e. A^{**} is the smallest monotone complete subset of A^{**} which contains A. For this we shall need some results on orthogonal states. We say two states ρ and ω on A are *orthogonal*, written $\rho \perp \omega$, if $\|\rho-\omega\|=2$. We shall identify a state ρ on A with its normal (and weak-$*$ continuous) extension $\bar{\rho}$ on A^{**}.

4.4.6. Lemma. *Let ρ and ω be states on a* JB *algebra A. Then $\rho \perp \omega$ if and only if there is a projection $p \in A^{**}$ such that $\rho(p)=\omega(1-p)=1$.*

Proof. If p is a projection in A^{**} such that $\rho(p)=\omega(1-p)=1$ let $s=2p-1$. Then s is a symmetry such that $(\rho-\omega)(s)=2$, hence $\|\rho-\omega\|=2$ (1.1.22).

Conversely assume $\rho \perp \omega$. Since by the Alaoglu theorem (1.1.17) the unit ball A_1^{**} of A^{**} is a weak-$*$ compact convex set there exists by the Krein–Milman theorem (1.1.6) an extreme point s of A_1^{**} such that $(\rho-\omega)(s)=2$. By the spectral theorem (3.2.4) it is easy to see that an extreme point of A_1^{**} must have its spectrum contained in $\{\pm 1\}$. Therefore s is a symmetry, hence there is a projection p in A^{**} such that $2=[\rho(2p)-1]-[\omega(2p)-1]$, or $\rho(p)=1+\omega(p)$. Since $0 \leqslant \rho(p) \leqslant 1$ and $0 \leqslant \omega(p) \leqslant 1$ this shows $\rho(p)=1=\omega(1-p)$.

4.4.7. If $B \subset A^{**}$ we denote by B_σ (resp. B_δ) the set of weak-$*$ limits of increasing (resp. decreasing) sequences in B. We write $B_{\sigma\delta}$ for $(B_\sigma)_\delta$. As before we denote by A_1 the unit ball of A.

4.4.8. Lemma. *Let ρ and ω be orthogonal states of a* JB *algebra A. Then there exists $y \in (A_1^+)_{\sigma\delta}$ such that $\rho(y)=1$, $\omega(y)=0$.*

Proof. We consider A as an ideal in $\tilde{A}$. By 1.2.8 A_1^+ is weak-$*$ dense in $A_1^{**} \cap A_+^{**}$. By 4.4.6 there exists a sequence (x_n) in A_1^+ such that

$$\rho(x_n) > \frac{n-1}{n} \quad \text{and} \quad \omega(x_n) < \frac{1}{n} 2^{-n}.$$

For $n \leqslant m$ let

$$y_{nm} = \left(1+\sum_{k=n}^{m} kx_k\right)^{-1} \circ \sum_{k=n}^{m} kx_k = f\left(\sum_{k=n}^{m} kx_k\right),$$

where f is the real function $f(t)=t/(1+t)=1-1/(1+t)$. By 3.5.3 f is operator monotone, i.e. $0\leq a\leq b$ implies $f(a)\leq f(b)$. Thus (y_{nm}) increases with m and decreases with n. Note also that $y_{nm}\in A$ since A is an ideal in $\tilde{A}$. Let

$$y_n=\text{weak-}*\lim_m y_{nm}\in(A_1^+)_\sigma.$$

Since (y_{nm}) decreases with n, (y_n) is a decreasing sequence. Let

$$y=\text{weak-}*\lim_n y_n\in(A_1^+)_{\sigma\delta}.$$

We have

$$\omega(y_{nm})\leq\omega\left(\sum_{k=n}^{m}kx_k\right)\leq\sum_{k=n}^{m}k\frac{1}{k}2^{-k}\leq 2^{-n+1}.$$

Since ω is normal on A^{**}, $0\leq\omega(y_n)\leq 2^{-n+1}$, hence $\omega(y)=0$. If $n\leq m$, since $0\leq x_n\leq 1$, we have

$$\rho(y_{nm})\geq\rho(y_{mm})=\rho((1+mx_m)^{-1}mx_m)\geq\rho\left(\frac{m}{m+1}x_m\right)$$

$$=\frac{m}{m+1}\rho(x_m)\geq\frac{m}{m+1}\frac{m-1}{m}=\frac{m-1}{m+1}.$$

Thus $\rho(y_n)\geq 1$, hence equal to 1. It follows that $\rho(y)=1$.

4.4.9. Let A be a JB algebra. We say a subspace B of A^{**} is *monotone closed* if $a_\alpha\in B$, $a_\alpha\nearrow a$ implies $a\in B$.

4.4.10. Theorem. *Let A be a JB algebra. Then A^{**} is the monotone completion of A, i.e. A^{**} is the smallest monotone closed subspace of A^{**} containing A.*

Proof. Since A^{**} is a JBW algebra (4.4.3) it suffices by 4.2.3 to show that each projection in A^{**} lies in the monotone completion of A. Let p be a projection in A^{**}, $p\neq 0,1$. We shall find, for any pair ρ, ω of states on A such that $\rho(p)=1$ and $\omega(p)=0$, a projection $p_{\rho,\omega}$ in the monotone completion of A satisfying $\rho(p_{\rho,\omega})=1$ and $\omega(p_{\rho,\omega})=0$.

Assume now that ρ, ω is such a pair. By 4.4.6 $\rho\perp\omega$ so there is by 4.4.8 $y\in(A_1^+)_{\sigma\delta}$ such that $\rho(y)=1$, $\omega(y)=0$. By 4.2.6 there is an increasing sequence (f_n) of continuous real functions with $f_n(0)=0$ such that $(f_n(y))$ converges to the range projection $r(y)$ of y. In particular $r(y)\in(A_1^+)_{\sigma\delta\sigma}$. Since $0\leq y\leq 1$, $1=\rho(y)\leq\rho(r(y))\leq 1$, so that $\rho(r(y))=1$. Since $\omega(y)=0$, the Cauchy–Schwarz inequality (3.6.2) implies $\omega(y\circ p(y))=0$ for all polynomials p. Since $f_n(0)=0$, then the Stone–Weierstrass theorem implies $\omega(f_n(y))=0$. By the normality of ω on A^{**} (4.4.3), $\omega(r(y))=0$. We can therefore let $p_{\rho,\omega}=r(y)$.

To convince ourselves that there are ρ, ω in $S(A)$ with the properties $\rho(p)=1$, $\omega(p)=0$, we shall show that more generally, if q is a nonzero projection in A^{**}, there is $\rho\in S(A)$ such that $\rho(q)=1$. Indeed, pick some $\rho_0\in S(A)$ such that $\rho_0(q)\neq 0$. Let $\rho=\rho_0(q)^{-1}U_q^*\rho_0$. Then ρ is a state on A^{**} such that $\rho(q)=1$. By 4.4.3 ρ is weak-$*$ continuous, and so belongs to $S(A)$.

Let $T=\{\phi\in S(A)\colon \phi(p)=0\}$. By 4.2.8 and its proof $p_\rho=\bigwedge_{\omega\in T}p_{\rho,\omega}$ belongs to the monotone completion of A, and if $\omega_1,\ldots,\omega_n\in T$ then $\bigvee_{i=1}^n p_{\rho,\omega_i}^\perp=(\bigwedge_{i=1}^n p_{\rho,\omega_i})^\perp$. Since $\rho(p_{\rho,\omega}^\perp)=0$ for all $\omega\in T$, $\rho(\sum_{i=1}^n p_{\rho,\omega_i}^\perp)=0$, hence it follows as for $r(y)$ above that $\rho(r(\sum_{i=1}^n p_{\rho,\omega_i}^\perp))=0$. But then by 4.2.8 $\rho(\bigvee_{i=1}^n p_{\rho,\omega_i}^\perp)=0$, and so $\rho(\bigwedge_{i=1}^n p_{\rho,\omega_i})=1$ whenever $\omega_1,\ldots,\omega_n\in T$. By normality of ρ and the proof of 4.2.8 $\rho(p_\rho)=1$. Since $\omega(p_{\rho,\omega})=0$ for all $\omega\in T$ it is clear that $\omega(p_\rho)=0$.

Let $S=\{\rho\in S(A)\colon \rho(p)=1\}$, and let $\tilde p=\bigvee_{\rho\in S}p_\rho$. We shall show $\tilde p=p$, thus proving that p belongs to the monotone completion of A, since it is clear by 4.2.8 that $\tilde p$ belongs to this set. As above $\rho(\tilde p)=1$ and $\omega(\tilde p)=0$ for $\rho\in S, \omega\in T$. The previous arguments applied to $\tilde p\wedge p$ now show $\rho(\tilde p\wedge p)=1$ and $\omega(p-\tilde p\wedge p)=0$ for all $\rho\in S$. But if $p\neq\tilde p\wedge p$ then there is $\rho\in S(A)$ such that $\rho(p-\tilde p\wedge p)=1$. But then $\rho\in S$, and we have a contradiction, hence $p=\tilde p\wedge p$, and $p\leq\tilde p$. If $\tilde p\neq p$ we can similarly find $\omega\in S(A)$ such that $\omega(\tilde p-p)=1$. But then $\omega\in T$, again a contradiction. Thus $p=\tilde p$, as asserted, and we have shown each projection in A^{**} different from 0 and 1 belongs to the monotone completion of A. Since evidently $1=\bigvee_{a\in A}r(a)$, the same is true for 1, and we are done.

4.4.11. Corollary. *Let A be a* JB *algebra and ρ a normal bounded linear functional on A^{**}. Then ρ is weak-$*$ continuous.*

Proof. By 4.4.10 ρ is uniquely determined by its restriction to A, and by 1.1.22 each functional in A^* has a unique weak-$*$ continuous extension to a functional on A^{**}. Since this extension is normal, it must be ρ, so ρ is weak-$*$ continuous.

4.4.12. A linear map $E\colon A\to A$ with A a JB algebra is said to be *idempotent* if $E^2=E$. In C^* algebras idempotent maps which map A_+ into A_+ have been quite important. In the JB algebra case we shall need the following result.

4.4.13. Lemma. *Let A be a* JB *algebra and B a* JB *subalgebra of A. Suppose $E\colon A\to B$ is a surjective linear idempotent map with $\|E\|\leq 1$. Then $E(a\circ b)=E(a)\circ b$ for all $a\in A, b\in B$.*

Proof. By 1.1.21 E extends to a map $E^{**}: A^{**} \to B^{**}$ such that $\|E^{**}\| \leq 1$ and E^{**} is continuous in the weak-$*$ topologies. By 1.1.23 we may consider B^{**} as a subset of A^{**} and then by weak-$*$ density of B in B^{**} (1.1.19) E^{**} is idempotent, hence surjective. By 4.4.3 A^{**} and B^{**} are JBW algebras with B^{**} a JBW subalgebra of A^{**}. Thus we may replace A, B and E by A^{**}, B^{**} and E^{**} respectively, i.e. by 4.1.7 and 4.2.3 we may assume

(i) A has an identity 1,
(ii) B has an identity e,
(iii) B is the norm-closed linear span of its projections.

We first show $E(1) = e$. Indeed, if $E(1-e) = b \neq 0$ let $\beta = \|b\|$. Then $b \in U_e(A)$, so that

$$\|1-e+\beta^{-1}b\| = \max\{\|1-e\|, \beta^{-1}\|b\|\} = 1.$$

Since $\|E(1-e+\beta^{-1}b)\| = \|b+\beta^{-1}b\| = \beta(1+\beta^{-1}) > 1$ we have contradicted the assumption that $\|E\| \leq 1$, proving our assertion.

Just as for states (1.2.2) it now follows that E is positive, i.e. $a \geq 0$ implies $E(a) \geq 0$. Indeed if $a \in A_1^+$ then $\|1-a\| \leq 1$, hence $\|e-E(a)\| = \|E(1-a)\| \leq 1$, so that $E(a) \geq 0$ by the spectral theorem (3.2.4) and the fact that e is the identity in B.

Let p be a projection in B. If $a \in A^+$ then by positivity of U_p (3.3.6) $0 \leq \{pap\} \leq \|a\|\, p$, so that $0 \leq E(\{pap\}) \leq \|a\|\, E(p) = \|a\|\, p$, and in particular by 4.1.13 $E(\{pap\}) \in U_p(A)$. By linearity this holds for all $a \in A$, so we have

$$\{pE(\{pap\})p\} = E(\{pap\}) \qquad \text{for all } a \in A, \tag{4.4}$$

or $U_pEU_p = EU_p$.

We first assume $e = 1$. Let $A = A_1 \oplus A_{1/2} \oplus A_0$ be the Peirce decomposition of A with respect to p (2.6.2). By (4.4) $E(A_1) \subset A_1$, and similarly $E(A_0) \subset A_0$. Let $a \in A_{1/2}$. If $\rho \in S(B)$ and $\rho(p) = 0$ then $\rho E \in S(A)$ and $\rho E(p) = \rho(p) = 0$. Since $p \circ a = \frac{1}{2}a$ we therefore have from the Cauchy–Schwarz inequality (3.6.2) that $\rho(E(a)) = 2\rho E(p \circ a) = 0$. For a general state $\rho \in S(B)$, $\rho U_{p^\perp}(p) = 0$, so that $\rho(U_{p^\perp}E(a)) = 0$. Since the states of B separate points $U_{p^\perp}E(a) = 0$, and $E(a) \in A_1 \oplus A_{1/2}$. By symmetry $E(a) \in A_{1/2} \oplus A_0$, hence $E(a) \in A_{1/2}$. We have thus shown $E(A_i) \subset A_i$, $i = 0, \frac{1}{2}, 1$. But then $ET_p = T_pE$ or $E(p \circ a) = p \circ E(a)$. Since by (iii) B is the closed linear span of its projections $E(b \circ a) = b \circ E(a)$ for all $b \in B$, $a \in A$.

Assume next $e \neq 1$ and let p be a projection in B as above. Let $A = A_1 \oplus A_{1/2} \oplus A_0$ be the Peirce decomposition of A with respect to e. Then $A_1 \supset B$, so $E(A_1) = B$. By the previous paragraph we then have $E(b \circ a) = b \circ E(a)$ for $a \in A_1$, $b \in B$. Since $A_1 \circ A_{1/2} \subset A_{1/2}$ and $A_0 \circ A_1 = 0$ (2.6.3) we have $B \circ A_{1/2} \subset A_{1/2}$ and $B \circ A_0 = 0$. Therefore it suffices to show $E(A_{1/2}) = E(A_0) = 0$. Let $\rho \in S(A)$ and $a \in A_0$ or $a \in A_{1/2}$. Then $a = e^\perp \circ c$ with c equal to a if $a \in A_0$ and $c = 2a$ if $a \in A_{1/2}$. Since

$\rho E \in S(A)$ and $\rho E(e^{\perp}) = 0$ the Cauchy–Schwarz inequality (3.6.2) implies $\rho(E(a)) = \rho(E(e^{\perp} \circ c)) = 0$. Since this holds for all states ρ, we get $E(a) = 0$, proving the lemma.

4.4.14. If X is a Banach space a norm-closed subspace Y of X^* is a *predual* for X if $X = Y^*$ in the natural duality $(x, \rho) \to \rho(x)$ between X and X^* (1.1.24). If X has a predual we say X is a *Banach dual space*. If V is a subspace of X recall from 1.1.9 that the *polar* V^0 of V is the subspace of X^* defined by $V^0 = \{\rho \in X^*: \rho(x) = 0 \text{ for all } x \in V\}$.

One of the classical results in C^* algebras is due to Sakai (see 1.4.6), and states that a C^* algebra is ultra-weakly closed if and only if it is a Banach dual space. The analogous result is true for JB algebras. We first prove a preliminary result.

4.4.15. Lemma. *Let M be a JBW algebra. Let K be the set of normal states on M and V the linear span of K in M^*. Let $J = V^0$ be the polar of V in M^{**}. Then there exists a central projection e in M^{**} such that $J = (1-e)M^{**}$, and the map $a \to ea$ is an isomorphism of M onto eM^{**}.*

Proof. By 4.1.5 $T_a^*(V) \subset V$ for all $a \in M$. Thus if $x \in J = V^0$ and $P \in V$ we have $T_a^{**}(x)(\rho) = x(T_a^* \rho) = 0$, so that $T_a^{**}(J) \subset J$ for all $a \in M \subset M^{**}$. By separate weak-$*$ continuity of multiplication in M^{**} (4.4.3) and the fact that J is weak-$*$ closed, we conclude that $T_a(J) \subset J$ for all $a \in M^{**}$. Thus J is an ideal in M^{**}, so by 4.4.4 there is a central projection f in M^{**} such that $J = fM^{**}$.

Let $e = 1 - f$. Then e is a central projection in M^{**}, so that by 2.5.6 U_e is a homomorphism of M^{**} into itself. Since by 2.5.5 $\iota = U_e + U_f$, the kernel of U_e is J. Now the normal states separate M, so that $M \cap J = M \cap V^0 = \{0\}$. Thus U_e is one-to-one on M and so is an isometry of M into eM^{**} (3.4.3).

It remains to show $U_e(M) = eM^{**}$. For this we first show $U_e(M)$ is monotone closed in M^{**}. Suppose $(U_e a_\alpha)$ is a bounded increasing net in $U_e(M)$. Since by 4.4.4 eM^{**} is a JBW algebra with identity e there is $a \in M^{**}$ such that $U_e a_\alpha \nearrow a$. Since U_e is an isomorphism of M onto $U_e(M)$, it is an order isomorphism, so (a_α) is a bounded increasing net in M. Let b be its least upper bound. Then $a_\alpha \to b$ strongly. By 4.1.8 U_e is strongly continuous, so $a = U_e b \in U_e(M)$, proving that $U_e(M)$ is monotone closed.

Let $M' = \{b \in M^{**}: U_e b \in U_e(M)\}$. Then $M' \supset M$, and by the previous paragraph M' is monotone closed. Thus by 4.4.10 $M' = M^{**}$, and so $eM^{**} = U_e(M^{**}) = U_e(M)$, proving that U_e is an isomorphism of M onto eM^{**}.

4.4.16. Theorem. *Let M be a JB algebra. Then M is a JBW algebra if and only if M is a Banach dual space. In this case the predual is unique and consists of the normal linear functionals on M.*

Proof. Suppose M is a JBW algebra and use the notation of 4.4.15. Let W denote the norm closure of V. Then $W^0 = V^0$, so if $f = 1 - e$ then $fM^{**} = J = W^0$, and by 1.1.15 if $a \in M^{**}$, $\|a \mid W\| = \|a + W^0\|$. Now U_e is a homomorphism on M^{**} with kernel J, so by 3.4.4 $\|U_e a\| = \|a + J\|$. Combining the two equalities we then have $\|a \mid W\| = \|U_e a\| = \|a\|$ for $a \in eM^{**}$, so that the restriction map $a \to a \mid W$ is an isometry from eM^{**} onto W^*. Since U_e is an isometry of M onto eM^{**} by 4.4.15, the composition $a \to U_e a \to U_e a \mid W$ is an isometry of M onto W^*. If $\rho \in W$ is identified with its weak-$*$ continuous extension to M^{**} we have $\rho(f) = 0$ since $f \in V^0$. Since f is a central projection $\rho = \rho U_e$, so that $U_e^*(\rho) = \rho U_e = \rho$ for $\rho \in W$. But then using dual notation, $\langle U_e a, \rho \rangle = \langle a, U_e^* \rho \rangle = \langle a, \rho \rangle = \rho(a)$ for $a \in M$. Therefore M is the dual of W in the natural pairing. This proves that M is a Banach dual space and that one predual is the norm-closed linear span of its normal states.

Conversely assume M is a Banach dual space and that X is a norm-closed subspace of M^* such that $M = X^*$ in the natural pairing. The inclusion map $X \to M^*$ defines by 1.1.20 a linear map $E: M^{**} \to M = X^*$ of norm 1 such that if $a \in M^{**}$ and $\rho \in X$ then $\rho(E(a)) = a(\rho)$. Since this holds for all $\rho \in X$, $E(a) = a$ for $a \in M$, hence E is idempotent. It is immediate from its definition or from 1.1.20 that E is continuous if M^{**} is given the weak-$*$ topology and M the weak topology defined by X. In particular the null space J of E is weak-$*$ closed in M^{**}. By 4.4.13 if $a \in M^{**}$, $b \in M$ then $E(a \circ b) = E(a) \circ b$, so if in particular $a \in J$, then $E(a \circ b) = 0$. Since M is weak-$*$ dense in M^{**} (1.1.19) it follows that $E(a \circ b) = 0$ for all $a \in J, b \in M^{**}$, hence J is an ideal in M^{**}. By 4.4.4 there is a central projection f in M^{**} such that $J = fM^{**}$. Let $e = 1 - f$. Then $M^{**} = J \oplus eM^{**}$, and E is a linear isomorphism of eM^{**} onto M.

If $a \in M$ then $a = E(a) = E(fa + ea) = E(ea)$, hence the inverse to the linear isomorphism $E: eM^{**} \to M$ is the map U_e, which is a homomorphism (2.5.6), hence an isomorphism of M onto eM^{**}. In particular M is a JBW algebra by 4.4.4.

It remains to characterize the predual. We first note that in the notation above $J = X^0$. For, if $a \in J$ and $\rho \in X$ then $\rho(a) = \rho(E(a)) = 0$, while conversely if $a \in X^0$ then for all $\rho \in X$ $\rho(E(a)) = \rho(a) = 0$, hence $E(a) = 0$ since X separates the points of M. By the bipolar theorem (1.1.10) therefore $\rho \in M^*$ belongs to X if and only if $\rho \mid J = 0$. Since $J = (1-e)M^{**}$, we conclude that $\rho \in X$ if and only if $\rho(a) = \rho(ea)$ for all $a \in M^{**}$.

If $\rho \in X$ then ρ, belonging to M^*, is normal as a functional on M^{**}

(4.4.3), and hence on eM^{**}. Since $a \to ea$ is an isomorphism from M onto eM^{**} and $\rho(a) = \rho(ea)$ for all $a \in M$, ρ is normal on M.

Conversely, if $\rho \in M^*$ is normal on M, let $\tilde{\rho} = \rho E$ on M^{**}. Since E is an isomorphism from eM^{**} onto M, $\tilde{\rho}$ is normal on eM^{**}, and therefore on all of M^{**}, since it vanishes on $e^{\perp}M^{**}$. Since $\tilde{\rho} \mid M = \rho$, by 4.4.11 $\tilde{\rho} = \rho$ as functionals on M^{**}. Thus ρ vanishes on J, and hence belongs to X.

We have proved that the predual of M is unique and consists of the normal functionals. This finishes the proof. We note for future reference that in the latter part of the proof, we showed the following result:

4.4.17. Corollary. *Let M be a JBW algebra. Then there is a central projection e in M^{**} such that the predual of M is $T_e^*(M^*)$, and T_e is an isomorphism of M onto eM^{**}.*

4.5. Normal functionals and homomorphisms

4.5.1. In this section we present some of the easy consequences of the existence and uniqueness of preduals for JBW algebras (4.4.16). In particular, the space V used in Section 4.4 turns out to be identical with the predual of M (4.5.4). Then we shall investigate continuity of homomorphisms of JBW algebras.

4.5.2. If M is a JBW algebra we denote by M_* its predual, i.e. the set of normal bounded linear functionals on M. We write M_*^+ for the cone of positive functionals in M_*.

4.5.3. Proposition. *Let M be a JBW algebra and let $\rho \in M_*$. Then there are $\sigma, \tau \in M_*^+$ such that $\rho = \sigma - \tau$ and $\|\rho\| = \|\sigma\| + \|\tau\|$.*

Proof. Since M is an order unit space, by 1.2.6 there are $\sigma_0, \tau_0 \in M_+^*$ such that $\rho = \sigma_0 - \tau_0$ and $\|\rho\| = \|\sigma_0\| + \|\tau_0\|$. Let $e \in M^{**}$ be the central projection such that $M_* = T_e^*(M^*)$ (4.4.17). Let $\sigma = T_e^*(\sigma_0)$, $\tau = T_e^*(\tau_0)$. Then $\sigma, \tau \in M_*^+$ and $\rho = \sigma - \tau$. Also, $\|\sigma\| + \|\tau\| \leq \|\sigma_0\| + \|\tau_0\| = \|\rho\| \leq \|\sigma\| + \|\tau\|$.

4.5.4. Corollary. *Let M be a JBW algebra, and let $\rho \in M^*$. Then the following are equivalent:*

(i) $\rho \in M_*$.

(ii) *ρ is weakly continuous.*

(iii) *ρ is strongly continuous.*

In particular, the weak and $\sigma(M, M_*)$ topologies coincide on M.

Proof. If K is the set of normal states on M then $M_*^+ = \{\lambda\rho \colon \lambda \geq 0, \rho \in K\}$. Hence by 4.5.3 M_* equals the linear span V of K. Thus the $\sigma(M, M_*)$

topology is the weak or $\sigma(M, V)$ topology (4.1.3). This proves the equivalence of (i) and (ii). Trivially, (ii) implies (iii). Finally, if ρ is strongly continuous and $a_\alpha \nearrow a$ then $a_\alpha \to a$ strongly by 4.1.3, so $\rho(a_\alpha) \to \rho(a)$, or ρ is normal. Thus (iii) implies (i).

4.5.5. Corollary. *Let M be a JW algebra acting on a complex Hilbert space. Then the weak topology on M (considered as a JBW algebra) coincides with the ultraweak topology.*

Proof. If $\tilde{M}_*$ denotes the set of ultraweakly continuous linear functionals on M then by 1.4.4 $M \cong (\tilde{M}_*)^*$. By the uniqueness of the predual of M (4.4.16) $\tilde{M}_* = M_*$. Since the ultraweak topology on M is the $\sigma(M, \tilde{M}_*)$ topology, the proof is completed by an application of 4.5.4.

It might be added, in order to avoid confusion, that the topology we have called the strong topology in JBW algebras coincides with what is usually called the ultrastrong topology in the JW algebra case. The interested reader can easily verify this by using 4.5.4.

4.5.6. Let M, N be JBW algebras and let $\phi: M \to N$ be a homomorphism. We call ϕ *normal* if $a_\alpha \nearrow a$ in M implies $\phi(a_\alpha) \nearrow \phi(a)$. Note that this definition is in terms of order structure alone, and does not involve topology. Nevertheless, ϕ is normal if and only if it is weakly continuous.

Indeed, ϕ is weakly continuous if and only if ϕ^* maps N_* into M_*, i.e. if and only if $\rho \in N_*$, $a_\alpha \nearrow a$ in M implies $\rho(\phi(a_\alpha)) \to \rho(\phi(a))$. But $\rho(\phi(a_\alpha)) \to \rho(\phi(a))$ for all $\rho \in N_*$ if and only if $\phi(a_\alpha) \nearrow \phi(a)$ in N, so this latter condition is just normality if ϕ, and our claim is proved.

Clearly any isomorphism of JBW algebras, being an order isomorphism, is normal and hence weakly continuous.

4.5.7. Proposition. *Let A be a JB algebra, M a JBW algebra and $\phi: A \to M$ a homomorphism. Then there is a unique normal homomorphism $\bar{\phi}: A^{**} \to M$ extending ϕ.*

Proof. Since A^* is a predual for A^{**}, by 4.4.16 and 4.5.4 the weak topology on A^{**} is the weak-$*$ topology. Since A is weak-$*$ dense in A^{**} (1.1.19), the uniqueness of $\bar{\phi}$ follows from 4.5.6.

To show existence, note that $\phi^{**}: A^{**} \to M^{**}$ is weakly continuous, so by density of A in A^{**} it is a homomorphism. By 4.5.6 ϕ^{**} is normal. By 4.4.17, there is a central projection $e \in M^{**}$ such that $a \to ea$ is an isomorphism of M onto eM^{**}. Its inverse, being an isomorphism, is normal (4.5.6), so its extension E to a homomorphism $E: M^{**} \to M$ by $E \mid e^\perp M^{**} = 0$ (this is the E from the proof of 4.4.16) is normal. $\bar{\phi} = E\phi^{**}$ is the desired normal extension of ϕ.

4.5.8. Corollary. *Let A be a* JB *algebra and let J be a norm-closed ideal in A. Then each homomorphism of J into a* JBW *algebra M can be extended to a homomorphism of A into M.*

Proof. The weak closure of J in A^{**} is clearly an ideal, since A is weakly dense in A^{**} and multiplication is separately weakly continuous. By 4.3.6 there is a central projection e in A^{**} such that the weak closure of J equals eA^{**}. The second dual J^{**} of J can then be identified with eA^{**}; see 1.1.23. Then by 4.5.7, a homomorphism $\phi: J \to M$ can be extended to a normal homomorphism $\bar{\phi}: eA^{**} \to M$. We can then obtain the desired extension $A \to M$ of ϕ by $a \to \bar{\phi}(ea)$.

4.5.9. Let M be a JBW algebra and N a norm-closed subalgebra of M. If N is monotone closed it is clearly monotone complete, and every normal state on M restricts to a normal state on N. Hence N is a JBW algebra. We call a monotone closed and norm closed subalgebra of M a *JBW subalgebra*. The following result is a natural analogue of the characterization of JBW algebras as dual spaces.

4.5.10. Proposition. *A norm-closed subalgebra of a* JBW *algebra M is a* JBW *subalgebra if and only if it is weakly closed.*

Proof. Assume A is a JBW subalgebra. Let $\iota: A \to M$ be the inclusion map. By 1.1.23 $\iota^{**}(A^{**})$ is weakly closed in M^{**}. Therefore, in the terminology of the proof of 4.5.7, since the isomorphism $E: eM^{**} \to M$ is a homomorphism of weak topologies (4.5.6), $\bar{\iota}(A^{**}) = E(e\iota^{**}(A^{**}))$ is weakly closed in M.

Let $B = \{a \in A^{**}: \bar{\iota}(a) \in A\}$. Since $\bar{\iota}$ is normal and A is monotone closed in M, B is monotone closed in A^{**}. But B contains A, and A^{**} is the monotone completion of A (4.4.10), so B equals A^{**}. Thus $A = \bar{\iota}(A^{**})$, and A is weakly closed.

The converse, that a weakly closed subalgebra is monotone closed, is trivial, since monotone nets converge weakly (4.1.3).

4.5.11. Corollary. *Let M and N be* JBW *algebras and $\phi: M \to N$ a normal homomorphism. Then $\phi(M)$ is a* JBW *subalgebra of N.*

Proof. Let e be the support of ϕ, i.e. the central projection in M such that $e^{\perp}M$ is the kernel of ϕ (4.3.7). Then $\phi(eM) = \phi(M)$, and on eM ϕ is injective, hence an isometry. Thus the unit ball of $\phi(M)$, being the image of the unit ball of eM, is weakly compact by the weak continuity of ϕ and the Alaoglu theorem (1.1.17). Now the reader who knows the Krein–Šmullyan theorem can conclude that $\phi(M)$ is weakly closed. Since we do not want to assume this theorem, we prove instead that $\phi(M)$ is monotone closed. Let (a_α) be a bounded increasing net in $\phi(M)$. We may assume $\|a_\alpha\| \leq 1$. But then, since the unit ball of $\phi(M)$ is weakly closed, $a = \text{weak} \lim a_\alpha$ belongs to $\phi(M)$, as desired.

4.5.12. The Kaplansky density theorem. *Let A be a strongly dense subalgebra of a* JBW *algebra M. Then the unit ball of A is strongly dense in the unit ball of M.*

Proof. We may assume A is norm closed. Let $\iota: A \to M$ be the inclusion map. As in the proof of 4.5.11, $\bar{\iota}$ maps the unit ball of A^{**} onto that of M. But the unit ball of A is weakly dense in that of A^{**} (1.1.19) and $\bar{\iota}$ is weakly continuous, so the unit ball of A is weakly dense in that of M.

By 4.5.4 the weak and strong topologies have the same continuous linear functionals. Therefore a convex set has the same closure in the two topologies, as follows from the Hahn–Banach theorem (1.1.2). Thus the unit ball of A is strongly dense in that of M.

4.6. Factor representations of JB algebras

4.6.1. The centre Z of a JBW algebra M is clearly a JBW subalgebra of M. If Z consists of scalar multiples of the identity alone, M is called a *JBW factor.* Since Z is generated by its projections (4.2.3), M is a JBW factor if and only if it has no central projections other than 0 and 1.

Let A be a JB algebra and ϕ a homomorphism of A into a JBW algebra M. We call ϕ a *factor representation* if the weak closure $\overline{\phi(A)}$ of $\phi(A)$ in M is a JBW factor. Two factor representations ϕ_1 and ϕ_2 are called *equivalent* if there is an isomorphism ψ of $\overline{\phi_1(A)}$ onto $\overline{\phi_2(A)}$ such that $\psi\phi_1 = \phi_2$.

In this section we show that all factor representations of A can be realized in A^{**}, and that each pure state on A, in particular, leads to a factor representation of A. This corresponds to the irreducible GNS representation associated with a pure state of a C^* algebra.

4.6.2. Proposition. *Let A be a* JB *algebra. For each minimal central projection e in A^{**} there is a factor representation $\phi_e: A \to eA^{**}$ defined by $\phi_e(a) = ea$. Two such factor representations ϕ_e and ϕ_f are equivalent if and only if $e = f$, and each factor representation of A is equivalent to ϕ_e for some minimal central projection e in A^{**}.*

Proof. Let e be a central projection in A^{**}. Since $A^{**} = eA^{**} \oplus e^\perp A^{**}$, a projection f in eA^{**} is central in eA^{**} if and only if it is central in A^{**}. Thus eA^{**} is a JBW factor if and only if e is a minimal central projection.

Since A is weakly dense in A^{**}, $\phi_e(A) = eA$ is weakly dense in eA^{**}. Therefore ϕ_e is a factor representation whenever e is a minimal central projection in A^{**}. Suppose ϕ_e and ϕ_f are equivalent. Then there is an isomorphism $\psi: eA^{**} \to fA^{**}$ such that $\psi(ea) = fa$ for all $a \in A$. But ψ is normal (4.5.6) and A is weakly dense in A^{**}, so $\psi(ea) = fa$ for all $a \in A^{**}$. In particular, $a = e$ and $a = 1$ yield $fe = \psi(ee) = \psi(er) = f$. By symmetry $fe = e$, so $e = f$.

Finally, let $\phi: A \to M$ be a factor representation of A. Let $\bar{\phi}: A^{**} \to M$ be its normal extension (4.5.7), and let e be the support of $\bar{\phi}$ in A^{**} (4.3.7). Then the restriction ψ of $\bar{\phi}$ to eA^{**} is an isomorphism of eA^{**} onto $\bar{\phi}(A^{**}) = \overline{\phi(A)}$, so eA^{**} is a factor, and e is a minimal central projection. If $a \in A$ then $\phi(a) = \psi(ea)$, so $\phi = \psi\phi_e$, and ϕ is equivalent to ϕ_e.

4.6.3. Let M be a JBW algebra and ρ a normal state on M. Then there is a smallest central projection e such that $\rho(e) = 1$. This projection is called the *central support* of ρ and is denoted by $e(\rho)$. If ρ is a state on a JB algebra A then it is a normal state on A^{**}, so by the above definition it has a central support $c(\rho)$ in A^{**}. We write $A_\rho = c(\rho)A^{**}$, and let $\phi_\rho: A \to A_\rho$ be the homomorphism $a \to c(\rho)a$.

4.6.4. Proposition. *If A is a JB algebra and ρ is a pure state on A then ϕ_ρ is a factor representation. For each $a \in A$ there is some pure state ρ such that $\|\phi_\rho(a)\| = \|a\|$.*

Proof. In order to show that ϕ_ρ is a factor representation we must by 4.6.2 prove that $c(\rho)$ is a minimal central projection. Assume on the contrary that e is a central projection in A^{**}, $0 < e < c(\rho)$. By definition of $c(\rho)$ (4.6.3), $\rho(e) < 1$ and $\rho(c(\rho) - e) < 1$, i.e. $0 < \rho(e) < 1$. Then we can write ρ as a convex combination $\rho = \lambda\sigma + (1-\lambda)\tau$, where $\lambda = \rho(e)$, $\sigma(a) = \lambda^{-1}\rho(ea)$, $\tau(a) = (1-\lambda)^{-1}\rho(e^\perp a)$. Since ρ is pure, $\sigma = \tau = \rho$, and in particular $\rho(e) = \sigma(e) = 1$, which is impossible. Thus $c(\rho)$ is minimal, and ϕ_ρ is a factor representation.

Next, let $a \in A$. By 3.6.8 there is a pure state ρ on A such that $|\rho(a)| = \|a\|$. Since $\rho(c(\rho)^\perp) = 0$, it follows from 3.6.3 that $\rho(a) = \rho(c(\rho)a)$. Thus $\|\phi_\rho(a)\| \geq |\rho(a)| = \|a\|$. The opposite inequality is obvious.

4.7. JC algebras

4.7.1. One of our main goals will be to describe how JC and JW algebras can be sorted out among JB and JBW algebras. In this connection it is important to know that certain JB and JBW algebras automatically are JC and JW algebras. In this section we prove some results to this effect. First, however, we need to know about the second dual of a C^* algebra.

4.7.2. Let $\mathscr{A}$ be a unital C^* algebra. For each state ρ on $\mathscr{A}_{sa}$, let $(\pi_\rho, H_\rho, \xi_\rho)$ be the associated GNS representation (1.3.10–11). Let H be the Hilbert space direct sum of all the H_ρ, and let $\pi = \bigoplus \pi_\rho : \mathscr{A} \to B(H)$ be the representation such that $\pi(a)\xi = \pi_\rho(a)\xi$ $(a \in \mathscr{A}, \xi \in H_\rho)$. Since $\mathscr{A}_{sa}$ is a unital JB algebra for each $a \in \mathscr{A}_{sa}$ there is a state ρ on $\mathscr{A}_{sa}$ such that $|\rho(a)| = \|a\|$ (3.6.8). Thus $|(\pi_\rho(a)\xi_\rho \mid \xi_\rho)| = \|a\|$, so $\|\pi_\rho(a)\| \geq \|a\|$, and $\|\pi(a)\| \geq \|a\|$. By 3.4.3 then $\|\pi(a)\| = \|a\|$. Using the C^* identity $\|x\|^2 = \|x^*x\|$ we conclude $\|\pi(x)\| = \|x\|$ for any $x \in \mathscr{A}$. π is known as the *universal representation* of $\mathscr{A}$. We shall now identify $\mathscr{A}$ with $\pi(\mathscr{A}) \subseteq B(H)$. The crucial property of this representation is that any state on $\mathscr{A}$ by construction is a vector state, i.e. of the form $a \to (a\xi, \xi)$ for some unit vector $\xi \in H$.

As in 1.4.2 any bounded linear functional on $\mathscr{A}$ is a linear combination of two bounded linear functionals which are real on $\mathscr{A}_{sa}$. Then, since $\mathscr{A}_{sa}$ is an order unit space, it follows from 1.2.6 that any bounded linear functional on $\mathscr{A}$ is a linear combination of four states, hence by the previous paragraph can be extended to an ultraweakly continuous linear functional on $B(H)$ (1.4.1).

We claim that the ultraweak closure $\bar{\mathscr{A}}$ of $\mathscr{A}$ in $B(H)$ is a von Neumann algebra which can be identified with $\mathscr{A}^{**}$.

That $\bar{\mathscr{A}}$ is a $*$-subalgebra of $B(H)$ is easy since $x \to x^*$ is ultraweakly continuous and multiplication is separately ultraweakly continuous. Recall from 1.4.4 that $\bar{\mathscr{A}}$ is a von Neumann algebra with predual $\bar{\mathscr{A}}_*$ consisting of all restrictions to $\bar{\mathscr{A}}$ of ultraweakly continuous linear functionals. Then the restriction map $\bar{\mathscr{A}}_* \to \mathscr{A}^*$ is linear, injective, has norm 1, and is also surjective since any bounded linear functional on $\mathscr{A}$ has an ultraweakly continuous extension to $B(H)$.

The map $\alpha : \mathscr{A}^{**} \to \bar{\mathscr{A}} = (\bar{\mathscr{A}}_*)^*$ dual to the restriction map is therefore easily seen to be linear, injective and a weak-$*$ to ultraweak homomorphism onto its image, and to extend the inclusion map $\mathscr{A} \to \bar{\mathscr{A}}$.

To show that α is onto $\bar{\mathscr{A}}$, note that as in 1.4.2 $(\mathscr{A}_{sa})^{**} = (\mathscr{A}^{**})_{sa}$. Since $\mathscr{A}_{sa}$ is a JB algebra and α is weak-$*$ to ultraweak continuous, α is a homomorphism of $\mathscr{A}_{sa}^{**}$ into $\bar{\mathscr{A}}_{sa}$, which is then onto by 4.5.11 and 4.5.5. Thus α maps $\mathscr{A}^{**} = \mathscr{A}^{**} + i\mathscr{A}^{**}$ onto $\bar{\mathscr{A}} = \bar{\mathscr{A}}_{sa} + i\bar{\mathscr{A}}_{sa}$.

We thus get a product on $\mathscr{A}^{**}$ making α an algebra $*$-isomorphism.

4.7.3. Lemma. *If* A *is a* JC *algebra then* A^{**} *is a* JW *algebra.*

Proof. We may assume $A \subseteq \mathscr{A}_{sa}$ for some unital C^* algebra $\mathscr{A}$. By 4.7.2 $\mathscr{A}^{**}$ is equipped with a separately weak-$*$ continuous product extending that of $\mathscr{A}$ such that $\mathscr{A}^{**}$ (with the weak-$*$ topology) is $*$-isomorphic and homomorphic to an ultraweakly closed $*$ subalgebra of $B(H)$ (with the ultraweak topology). Applying 1.1.22 to the inclusion map $A \to \mathscr{A}_{sa}$ we see that A^{**} can be identified with a weak-$*$ closed subalgebra of $\mathscr{A}^{**}$, hence with an ultraweakly closed subalgebra of $\bar{\mathscr{A}} \subseteq B(H)$. Thus A^{**} is a JW algebra.

4.7.4. Proposition. *Let A be a* JC *algebra and J a norm-closed ideal in A. Then A/J is a* JC *algebra, and in particular each homomorphic image of A is a* JC *algebra.*

Proof. By 3.4.2 A/J is a JB algebra, so contained in some JBW algebra M. Let $\phi: A \to A/J$ be the quotient map, and let by 4.5.7 $\bar{\phi}: A^{**} \to M$ be the weak-$*$ to weak continuous extension of ϕ. Then $J = \ker \bar{\phi}$ is a weak-$*$ closed ideal in A^{**}, hence J is by 4.4.4 of the form $J = (1-f)A^{**}$ with f a central projection in A^{**}. Since the map $\phi(a) \to fa$ is an isometry of $\phi(A)$ into A^{**} (cf. 3.4.4), $\phi(A)$ is a JC algebra by 4.7.3.

4.7.5. Proposition. *Suppose M is both a* JBW *algebra and a* JC *algebra. Then M is a* JW *algebra.*

Proof. Let ϕ be an isomorphism of M into the self-adjoint part of a C^* algebra $\mathscr{A}$. By 1.1.22 ϕ extends to a weak-$*$ to weak-$*$ continuous isometry $\phi^{**}: M^{**} \to \mathscr{A}^{**}$, which by separate weak-$*$ continuity of multiplication and weak-$*$ density of M in M^{**} is an isomorphism. By 4.4.17 there is a central projection e in M^{**} such that $U_e: M \to eM^{**}$ is a surjective isomorphism, and by 4.1.6 and 4.5.4 U_e is weak to weak-$*$ continuous. Thus $\phi^{**}U_e$ is a weak to weak-$*$ continuous isomorphism of M onto $\phi^{**}(e)\phi^{**}(M^{**})$. By 1.1.22 $\phi^{**}(M^{**})$ is weak-$*$ closed in $\mathscr{A}^{**}$, hence by 4.5.4 $\phi^{**}(e)\phi^{**}(M^{**})$ is a JBW subalgebra of the self-adjoint part of $\mathscr{A}^{**}$. Since $\mathscr{A}^{**}$ is a von Neumann algebra (4.7.2), $\phi^{**}U_e(M)$ is a JB algebra.

4.8. Comments

JW algebras were first studied by Topping [110] and Størmer [103], while the study of JBW algebras was initiated by Alfsen, Shultz and Størmer [19] even though preliminary work in this direction had been done by Janssen [64–67]. In Alfsen *et al.* [19] only that part of the theory which was needed to prove the structure theorem for JB algebras (7.2.3 below) was developed. As a result many theorems appeared in a weaker form than are presented here. The stronger results are mainly due to Shultz

[91]. For example, Section 4.1 is a blend of the contents in Alfsen *et al.* [19] and Shultz [91]. In the former [19] it was not shown that the second dual of a JB algebra is a JBW algebra, only a weaker result was proved. Shultz [91] succeeded in proving the general result. The proof we have presented is due to Hanche-Olsen [55]. The assertion (4.4.10) that A^{**} is the monotone completion of A for A a JB algebra is a direct Jordanification of the same theorem and proof for C^* algebras due to Pedersen [82].

The characterization of JBW algebras as Banach dual spaces is due to Shultz [91]. Our proof differs from his in our use of idempotent maps (4.4.13). The results in Section 4.5 can be found in one form or other in the literature, and some are 'folklore'. Section 4.6 is mainly taken from Alfsen *et al.* [19], while the first two results of Section 4.7 are due to Effros and Størmer [41].

Just as in Chapter 3 we have said much less about state spaces than what is known. Iochum and Shultz [60] have characterized the norm-closed convex set of normal states of both JBW and von Neumann algebras among all convex sets. Closely related to the normal states is a certain self-dual cone, which is the Jordan analogue of the natural cone associated with a faithful normal state on a von Neumann algebra. This cone has been studied by Bellisard and Iochum [27, and references therein]. In von Neumann algebras the natural cone is closely related to the so-called Tomita–Takesaki theory. The Jordan analogue of this theory has been studied by Haagerup and Hanche-Olsen [53].

5

Dimension theory

5.1. The projection lattice

5.1.1. In the theory of von Neumann algebras one of the basic techniques is that of comparison of projections, usually called dimension theory. Two projections p and q in a von Neumann algebra $\mathcal{M}$ are said to be equivalent if there exists an element v in $\mathcal{M}$ such that $v^*v = p$, $vv^* = q$. A partial ordering is introduced on the projection lattice, c.f. 4.2.7, by saying $p \lesssim q$ if p is equivalent to a subprojection of q. In analogy with cardinal numbers one then defines finite and infinite projections. If we call an additive homogeneous function $\tau: \mathcal{M}^+ \to \mathbb{R}^+ \cup \{+\infty\}$ a *trace* provided $\tau(uxu^*) = \tau(x)$ for all unitaries u in $\mathcal{M}$, then a trace is constant on each equivalence class of projections. Thus projections, in the presence of a trace, have a natural dimension, hence the name 'dimension theory'.

We shall in the present chapter generalize this theory to JBW algebras. Since we have no nonself-adjoint elements in a JBW algebra the definition of equivalence has to be modified, and the theory turns out to be closer to lattice theory than is the case for von Neumann algebras. In particular, some of the proofs are quite different from the corresponding ones in von Neumann algebra theory.

5.1.2. By a *lattice* we shall mean a partially ordered set L such that if $e, f \in L$ the set $\{g \in L: g \geq e, g \geq f\}$ has a smallest element $e \vee f$, and similarly there is a largest element $e \wedge f$ smaller than or equal to both e and f. We assume there is a smallest element $0 \in L$ and a largest element $1 \in L$. A subset K of L is called a *sublattice* if $e \vee f$ and $e \wedge f$ belong to K whenever e and f do.

Let L be a lattice. We say L is *modular* if $e \leq g$ implies

$$(e \vee f) \wedge g = e \vee (f \wedge g), \qquad f \in L.$$

If $e \in L$ we say an element $e' \in L$ is a *complement* for e if it satisfies $e \wedge e' = 0$ and $e \vee e' = 1$. Two elements e and f in L are said to be *perspective* if they have a common complement.

L is called *orthocomplemented* if there is given an order reversing map $e \to e^{\perp}$ on L satisfying $e^{\perp\perp} = e$ such that $e^{\perp}$ is a complement for e. L is called *orthomodular* if in addition $e \leq f$ implies

$$f = e \vee (f \wedge e^{\perp}).$$

Clearly a modular, orthocomplemented lattice is orthomodular. The converse is false.

It is sometimes illustrative to draw pictures to describe lattices. If two elements e and f are comparable then one draws a line from e to f such that if $e \leq f$ then f is placed higher up than e. e and f are not comparable if there is no connecting set of lines all going up or down from e to f.

5.1.3. Proposition. *Let L be an orthomodular lattice. Then the following are equivalent:*

(i) *L is modular.*

(ii) *If e and f are perspective elements in L with $e \leq f$ then $e = f$.*

(iii) *L contains no sublattice containing* 0 *and* 1 *and isomorphic to the lattice*

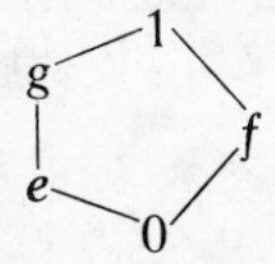

Proof. In the lattice described in (iii) e and g are perspective with common complement f. Moreover, this lattice is nonmodular. It is thus immediate that (ii)$\Leftrightarrow$(iii) and (i)$\Rightarrow$(iii).

We show (ii)$\Rightarrow$(i). Let $e,f,g \in L$ with $e \leq g$. In order to show that L is modular it suffices to show $(e \vee f) \wedge g \leq e \vee (f \wedge g)$, since the opposite inequality is always true.

If $p \in L$ we denote by $[0, p]$ the sublattice consisting of $q \in L$ such that $q \leq p$. With orthocomplementation $q \to p \wedge q^{\perp}$ this sublattice is clearly orthomodular. Hence we may assume that $e \vee f = 1$. We thus have $g \geq e \vee (f \wedge g)$ and must show the opposite inequality. To do this we show g and $e \vee (f \wedge g)$ are perspective with common complement $f \wedge (f \wedge g)^{\perp}$ and then apply (ii). Now it is immediate that if p, q, $r \in L$ then $(p \vee q) \vee r = p \vee (q \vee r)$ and similarly for $\wedge$. Thus we have

$$g \wedge (f \wedge (f \wedge g)^{\perp}) = (g \wedge f) \wedge (f \wedge g)^{\perp} = 0.$$

If we apply orthomodularity to $f \wedge g \leq f$ we have

$$(e \vee (f \wedge g)) \vee (f \wedge (f \wedge g)^{\perp}) = e \vee ((f \wedge g) \vee (f \wedge (f \wedge g)^{\perp})) = e \vee f = 1,$$

which proves the assertion.

5.1.4. Let M be a JBW algebra. We denote by P_M or simply P the lattice of projections in M. Z_M, or simply Z, will denote the centre of M, and $P_Z = P \cap Z$ is the set of central projections in M. We write $p^\perp = 1 - p$ if $p \in P$, and note that P is then an orthocomplemented lattice, since clearly $p \leqslant q$ implies $q \wedge p^\perp = q - p$ and so $q = p \vee (q \wedge p^\perp)$. As before the range projection of $a \in M$ will be denoted by $r(a)$ and the central support of $p \in P$ by $c(p)$.

If $s \in M$ is a symmetry, i.e. $s^2 = 1$, then U_s is an automorphism of M since by Macdonald's theorem $(U_s a) \circ (U_s b) = U_s(\{as^2b\}) = U_s\{a \circ b\}$. These automorphisms generate a group Int M, called the group of *inner automorphisms* of M. Notice that by 4.3.2 Z is the fixed point set of this group.

Two projections p and q in M are called *equivalent* if there is $\alpha \in \text{Int}\, M$ such that $q = \alpha(p)$. We then write $p \sim q$. If α can be written as $\alpha = U_{s_1} U_{s_2} \ldots U_{s_n}$ we write $p \underset{n}{\sim} q$. If $n = 1$ we say p and q are *exchanged by a symmetry*. It should be noted that if $p \sim q$ then $p^\perp \sim q^\perp$ in contrast with the equivalence of projections in a von Neumann algebra. Also we note that $\{q \in P : q \sim p\}$ is Int M invariant, hence the same is true for the supremum of this family, which must therefore be central. Since conversely central projections are not moved by inner automorphisms, we conclude

$$c(p) = \bigvee \{q \in P : q \sim p\}.$$

A projection $p \in M$ is called *Abelian* if M_p, cf. 4.1.13, is associative; p is *modular* if the projection lattice $[0, p]$ of M_p is modular. If 1 is modular, M itself is called modular. The set of Abelian and modular projections are Int M invariant, so we can define central projections e_{I} and e_{III} in M by

$$e_{\mathrm{I}} = \bigvee \{p \in M : p \text{ is Abelian}\},$$

$$e_{\mathrm{III}}^\perp = \bigvee \{p \in P : p \text{ is modular}\}.$$

Clearly $e_{\mathrm{I}} \leqslant e_{\mathrm{III}}^\perp$. Let $e_{\mathrm{II}} = 1 - e_{\mathrm{I}} - e_{\mathrm{III}}$. M is said to be of *type* I (resp. II, III) if $e_{\mathrm{I}} = 1$ (resp. $e_{\mathrm{II}} = 1$, $e_{\mathrm{III}} = 1$). If we select a maximal family (p_α) of Abelian projections with pairwise orthogonal central supports we find $\sum_\alpha c(p_\alpha) = e_{\mathrm{I}}$, and so $p = \sum_\alpha p_\alpha$ is Abelian with $c(p) = e_{\mathrm{I}}$. Similarly we can find a modular projection with central support e_{II}. We summarize these results in the following theorem.

5.1.5. Theorem. *Let M be a* JBW *algebra. Then M can be split uniquely into a direct sum of parts of type* I, II *and* III, *the different parts being characterized as follows*:

(i) *M is of type* I *if and only if there is an Abelian projection p in M with $c(p) = 1$.*

(ii) *M is of type* II *if and only if there is a modular projection p in M with $c(p)=1$, and M contains no nonzero Abelian projection.*
(iii) *M is of type* III *if and only if it contains no nonzero modular projection.*

5.1.6. Just as for von Neumann algebras we can define a finer decomposition. Any JBW algebra has a largest central modular projection. If this is 0, M is called *purely nonmodular*; if it is 1, M is modular. We say M is of type II_1 if it is modular and of type II, and it is of type II_∞ if it is purely nonmodular and of type II. Type I algebras split in the same way, but we shall obtain a finer decomposition for such algebras; see 5.3.5. Of course, if M is of type III, it is purely nonmodular.

5.2. Equivalence of projections

5.2.1. Lemma. *Let M be a* JBW *algebra with projection lattice P. If $p, q \in P$ there is a symmetry $s \in M$ such that $U_s\{pqp\}=\{qpq\}$.*

Proof. Let $a=p+q-1$. By the spectral theorem (3.2.4) and 4.2.6 there is $t \in M$, indeed $t \in W(a)$, such that $t^2=r(a)$ and $a=t\circ|a|$, where $|a|=(a^2)^{1/2}$ is the absolute value of a. Namely, let $t=r(a^+)-r(a^-)$. Let $s=t+1-r(a)$. Then s is a symmetry. We find $a^2=1-p-q+2p\circ q$. By the Shirshov–Cohn theorem (2.4.14), the algebra A generated by p, q and 1 is special. In an associative algebra $\mathcal{A}$ containing A as a Jordan subalgebra we can then write $a^2=1-p-q+pq+qp$, which implies $pa^2=pqp=a^2p$. Hence p and a^2 commute, so they generate a commutative subalgebra of $\mathcal{A}$ and hence an associative subalgebra of A. By 2.5.5 p operator commutes with a^2, and hence by 4.2.5 with $|a|$. From the above we also get $\{pqp\}=p\circ a^2$. Furthermore, since p and $|a|$ operator commute we have $p\circ|a|^2=\{|a|\ p\ |a|\}$. Just use the definition of the triple product to see this. From these two equalities we have $\{pqp\}=\{|a|\ p\ |a|\}$. Since $s \in W(a)$, s operator commutes with all elements of $W(a)$. Note also that then $U_s=2T_s^2-\iota$ commutes with $U_b=2T_b^2-T_{b^2}$ for all $b \in W(a)$. We therefore get, using 2.4.21, 2.4.18 and the identity $\{|a|^{1/2}s\ |a|^{1/2}\}=s\circ|a|=a$, which is true since s, $|a| \in W(a)$:

$$\begin{aligned}U_s\{pqp\}&=U_s\{|a|\ p\ |a|\}=U_sU^2_{|a|^{1/2}}p=U_{|a|^{1/2}}U_sU_{|a|^{1/2}}p\\&=U_{\{|a|^{1/2}s|a|^{1/2}\}}p=U_ap=\{(q-p^\perp)p(q-p^\perp)\}=\{qpq\}.\end{aligned}$$

5.2.2. Lemma. *Let M be a* JBW *algebra with projection lattice P. If $p, q \in P$ then the range projection of $\{pqp\}$ is given by $r(\{pqp\})=p-p\wedge q^\perp$.*

Proof. It is clear that $r(\{pqp\})\leq p$, hence there is $h \in P$ such that

$r(\{pqp\}) = p - h$. By 2.4.18 it follows that $0 = U_{\{pqp\}}h = U_pU_qU_ph = U_pU_qh$. Since $U_h = U_hU_p$ it follows from the identity $\{aba\}^2 = U_aU_ba^2$ (2.4.17) that $0 = U_hU_pU_qh = U_hU_qh = \{hqh\}^2$. Therefore by 4.2.2 $h \leqslant q^\perp$, so that $h \leqslant p \wedge q^\perp$.

On the other hand by strong continuity of U_y for $y \in M$ (4.1.8) and construction of $r(x)$, see proof of 4.2.6, if $x \geqslant 0$ and $U_yx = 0$ then $U_yr(x) = 0$. (Note that $0 \leqslant U_yx^n \leqslant \|x\|^{n-1}U_yx = 0$.) Since

$$U_{p\wedge q^\perp}\{pqp\} = U_{p\wedge q^\perp}U_pq = U_{p\wedge q^\perp}q = 0,$$

we therefore have $U_{p\wedge q^\perp}r(\{pqp\}) = 0$, and so $p \wedge q^\perp \leqslant r(\{pqp\})^\perp$. But then

$$h = p - r(\{pqp\}) = p \wedge r(\{pqp\})^\perp \geqslant p \wedge (p \wedge q^\perp) = p \wedge q^\perp,$$

showing that $h = p \wedge q^\perp$.

5.2.3. Proposition. *Let M be a* JBW *algebra with projection lattice P. If $p, q \in P$ we have:*

(i) $p - p \wedge q^\perp \underset{1}{\sim} q - q \wedge p^\perp$.
(ii) If $p \wedge q^\perp = q \wedge p^\perp = 0$ *then* $p \underset{1}{\sim} q$.
(iii) $(p \vee q) - q \underset{1}{\sim} p - (p \wedge q)$.
(iv) *If p and q are perspective then* $p \underset{2}{\sim} q$.

Proof. By 5.2.1, there is a symmetry s such that $U_s\{pqp\} = \{qpq\}$. Since U_s is an automorphism of M it preserves range projections, so $U_s(r(\{pqp\})) = r(\{qpq\})$. Thus (i) follows from 5.2.2. (ii) is immediate from (i), and so is (iii) by applying (i) to p and $q^\perp$, noticing that $q^\perp - q^\perp \wedge p^\perp = q^\perp - (p \vee q)^\perp = p \vee q - q$. To show (iv) note that if r is a complement of p then $p \wedge (r^\perp)^\perp = p \wedge r = 0$, and $p^\perp \wedge r^\perp = (p \vee r)^\perp = 0$, so that p and $r^\perp$ satisfy the conditions of (ii), whence $p \underset{1}{\sim} r^\perp$. If r is also a complement of q then $r^\perp \underset{1}{\sim} q$, and so $p \underset{2}{\sim} q$.

5.2.4. It is not hard to prove a partial converse to 5.2.3(iv), namely if $p, q \in P$ are exchanged by a symmetry then they are perspective.

We next arrive at the crucial lemma, which allows us to reduce many questions about $\sim$ to questions about $\underset{1}{\sim}$. We shall, however, first prove a Jordan identity which will be needed in the proof.

5.2.5. Lemma. *Let A be a Jordan algebra over $\mathbb{R}$ and $a, b, c \in A$. Then the following identity holds:*

$$4\{abc\}^2 = 4\{a\{b(a \circ c)b\}c\} - 2\{aba\} \circ \{cbc\} + \{a\{bc^2b\}a\} + \{c\{ba^2b\}c\}.$$

Proof. From the identity $\{aba\}^2 = \{a\{ba^2b\}a\}$ (2.4.17) we have

$$\{(a+c)b(a+c)\}^2 = \{(a+c)\{b(a+c)^2b\}(a+c)\}.$$

All terms of degree 2 in both a and c on both sides must coincide, as is easily seen by replacing a by λa and c by μc, $\lambda, \mu \in \mathbb{R}$. If we let $\equiv$ denote congruence modulo terms of degree $\neq 2$ in both a and c we have

$$\{(a+c)b(a+c)\}^2 = (\{aba\}+2\{abc\}+\{cbc\})^2$$
$$\equiv 4\{abc\}^2 + 2\{aba\}\circ\{cbc\}.$$

$$\{(a+c)\{b(a+c)^2b\}(a+c)\}$$
$$= \{(a+c)\{b(a^2+2a\circ c+c^2)b\}(a+c)\}$$
$$= \{(a+c)(\{ba^2b\}+2\{b(a\circ c)b\}+\{bc^2b\})(a+c)\}$$
$$\equiv 4\{a\{b(a\circ c)b\}c\}+\{a\{bc^2b\}a\}+\{c\{ba^2b\}c\}.$$

If we compare the two congruences we obtain the desired identity.

5.2.6. Lemma. *Let M be a* JBW *algebra with projection lattice P. Let $p, q \in P$ and suppose $p \perp q$ and $p \underset{2}{\sim} q$. Then $p \underset{1}{\sim} q$.*

Proof. The proof is quite simple in the case of JW algebras. In order to illustrate the ideas we first prove the lemma in that case, hence we first assume $M \subset B(H)_{\mathrm{sa}}$ is a JW algebra. By assumption there are symmetries $s, t \in M$ such that $q = \{s\{tpt\}s\}$. Let $x = pts \in B(H)$. Then $x^*x = q$, $xx^* = p$ and $x = xq = px$, so that $x^2 = xqpx = 0$. Therefore $(x+x^*)^2 = 2x\circ x^* = p+q$, and we have

$$\{(x+x^*)p(x+x^*)\} = x^*px = q,$$

$$\{(x+x^*)q(x+x^*)\} = xqx^* = p.$$

Since $x+x^* = 2\{pts\} \in M$, p and q are exchanged by the symmetry $x+x^*+1-p-q$ in M, proving the lemma for JW algebras.

We now return to the case of general JBW algebras. The above proof indicates what we shall do. As above we assume s and t are symmetries in M such that $q = \{s\{tpt\}s\}$, and we let

$$r = 2\{pts\}.$$

The proof is complete when we have shown

$$r^2 = p+q, \qquad \{rpr\} = q, \qquad \{rqr\} = p.$$

Let $M = \bigoplus_{1\leq i\leq j\leq 3} M_{ij}$ be the Peirce decomposition corresponding to p, q and $1-p-q$. The bulk of the proof consists of showing that $r \in M_{12}$ and $r^2 = p+q$. Indeed, if this is done then

$$\{rpr\} = 2r\circ(r\circ p) - r^2\circ p = r^2 - r^2\circ p = r^2\circ(1-p) = q,$$

and similarly $\{rqr\} = p$.

We first show $\{p^\perp rp^\perp\} = 0$. This will be a consequence of the following identity, which is a straightforward consequence of Macdonald's theorem (2.4.13)

$$\{(1-a)\{abc\}(1-a)\}=\{cb(a-a^2)\}-\{a\{(1-a)bc\}a\}, \qquad a, b, c \in M.$$

It follows that

$$\{(1-p)\{pts\}(1-p)\}=-\{p\{(1-p)ts\}p\}.$$

If $M=M_1 \oplus M_{1/2} \oplus M_0$ is the Peirce decomposition of M with respect to p we see that the left side belongs to M_0 while the right side belongs to M_1. Thus both sides are zero. In particular $\{p^\perp r p^\perp\}=0$, as asserted. It follows that $r \in M_{11} \oplus M_{12} \oplus M_{13}$.

We shall next show that

$$\{pts\}=\{tws\}=\{tsq\}, \tag{5.1}$$

where w is the projection

$$w=\{tpt\}=\{sqs\}. \tag{5.2}$$

It will then follow by the above argument applied to q instead of p that $r \in M_{12} \oplus M_{22} \oplus M_{23}$, so that r necessarily belongs to M_{12}.

By Macdonald's theorem the following identity is easily verified for $a, b, c \in M$.

$$\{a\{aba\}c\}=2a^2 \circ \{bac\}-2\{(a^2 \circ c)ab\}+\{ba^3c\}.$$

If we apply this identity we obtain

$$\begin{aligned}\{tws\}&=\{t\{tpt\}s\}\\ &=2t^2 \circ \{pts\}-2\{(t^2 \circ s)tp\}+\{pt^3s\}\\ &=2\{pts\}-2\{stp\}+\{pts\}\\ &=\{pts\}.\end{aligned}$$

By (5.2) and symmetry $\{tws\}=\{tsq\}$ proving (5.1). Thus we have shown

$$r=2\{pts\}=2\{tws\}=2\{tsq\} \in M_{12}.$$

In particular $r^2=4\{tws\}^2$, and 5.2.5 is applicable with $a=s$, $b=w$, $c=t$. If we note that $U_s w=q$, $U_t w=p$ we have

$$\begin{aligned}r^2&=4\{swt\}^2\\ &=4\{s\{w(s \circ t)w\}t\}-2\{sws\} \circ \{twt\}+U_sU_wt^2+U_tU_ws^2\\ &=4\{sU_w(s \circ t)t\}-0+U_sw+U_tw\\ &=4\{sU_w(s \circ t)t\}+q+p.\end{aligned}$$

Thus in order to show $r^2=p+q$ it suffices to show $U_w(s \circ t)=0$.

Since by Macdonald's theorem $\{t(s\circ t)t\}=\{tst^2\}$, we have $U_t(s\circ t)=\{ts1\}=s\circ t$. Therefore we have, using $U_{\{aba\}}=U_aU_bU_a$ (2.4.18),

$$\begin{aligned} U_w(s\circ t) &= \{\{tpt\}(s\circ t)\{tpt\}\} \\ &= U_tU_pU_t(s\circ t) \\ &= U_tU_p(s\circ t), \end{aligned}$$

hence it suffices to show $U_p(s\circ t)=0$. For this we use yet another identity, which is also an easy consequence of Macdonald's theorem, namely if a, b, $c\in M$ we have

$$\{a(b\circ c)a\}\circ a = a^2\circ\{abc\}+\tfrac{1}{2}\{a\{abc\}a\}-\tfrac{1}{2}\{a^3bc\}.$$

If we apply this and use $\{pts\}=\frac{1}{2}r\in M_{12}$ we find

$$\begin{aligned} U_p(s\circ t) &= \{p\{s\circ t\}p\}=\{p\{s\circ t\}p\}\circ p \\ &= p^2\circ\{pts\}+\tfrac{1}{2}\{p\{pts\}p\}-\tfrac{1}{2}\{p^3ts\} \\ &= \tfrac{1}{2}\{pts\}+0-\tfrac{1}{2}\{pts\} \\ &= 0. \end{aligned}$$

This completes the proof of the lemma.

5.2.7. Lemma. *Let M be a* JBW *algebra with projection lattice P. Suppose p and q in P are nonzero and equivalent. Then there are nonzero $p_1, q_1\in P$ such that $p_1\leqslant p, q_1\leqslant q$ and $p\underset{1}{\sim}q_1$.*

Proof. Assume $p\underset{n}{\sim}q$ (see 5.1.4). If $n=1$ there is nothing to prove, so we may by induction assume the lemma holds for $n-1$. We may assume $n>1$. Furthermore we may assume $p\perp q$, since otherwise we could choose $p_1=p-p\wedge q^\perp$, $q_1=q-q\wedge p^\perp$ and then apply 5.2.3. By definition of $\underset{n}{\sim}$ there is $r\in P$ such that $p\underset{n-1}{\sim}r\underset{1}{\sim}q$. By the induction hypothesis there are nonzero $p_1\leqslant p$ and $r_1\leqslant r$ in P such that $p_1\underset{1}{\sim}r_1$. If $q=\{srs\}$ let $q_1=\{sr_1s\}\leqslant q$. Then $p_1\underset{2}{\sim}q_1$, so by 5.2.6 $p_1\underset{1}{\sim}q_1$.

5.2.8. In order to simplify notation in the arguments to follow we introduce the analogue of partial isometries, a concept which was implicitly used in the proof of 5.2.6.

An element $s\in M$ is called a *partial symmetry* if s^2 is a projection. If p, $q\in P$ and s is a partial symmetry such that $s^2=p\vee q$ and $\{sps\}=q$ then $p\underset{1}{\sim}q$. Indeed $t=s+1-p\vee q$ is a symmetry such that $\{tpt\}=q$. Conversely, if t is a symmetry such that $\{tpt\}=q$ then $\{t(p\vee q)t\}=p\vee q$, so $s=t-1+p\vee q$ is a partial symmetry such that $s^2=p\vee q$ and $\{sps\}=q$.

Assume that $p\perp q$ and that s is a partial symmetry such that $\{sps\}=q$ and $s^2=p+q$. Then we find $s\circ q=s\circ\{sps\}=\{s^2ps\}=s\circ p$, and since

$s\circ(p+q)=s$, we conclude $s\circ p=s\circ q=\frac{1}{2}s$. Considering the Peirce decomposition induced by p and q, we conclude that $s\in\{pMq\}$. In other words, p and q are strongly connected.

5.2.9. Lemma. *Let M be a JBW algebra with projection lattice P, and let J be an index set. Let p, q, p_α, $q_\alpha\in P$, $\alpha\in J$, satisfy $p\perp q$, $p=\sum_{\alpha\in J}p_\alpha$, $q=\sum_{\alpha\in J}q_\alpha$ and $p_\alpha\,\underset{\top}{\sim}\,q_\alpha$ for all $\alpha\in J$. Then $p\,\underset{\top}{\sim}\,q$.*

Proof. Let s_α be a partial symmetry such that $s_\alpha^2=p_\alpha+q_\alpha$ and $q_\alpha=\{s_\alpha p_\alpha s_\alpha\}$, $\alpha\in J$. Let $s=\sum_{\alpha\in J}s_\alpha$. Then $s^2=p+q$ and $\{sps\}=q$.

5.2.10. Lemma. *Let M be a JBW algebra with projection lattice P. Let p_i, $q_i\in P$, $i=1,2$, satisfy $p_1+p_2=p\in P$, $q_1+q_2=q\in P$, $p_1\perp q_2$, $p_2\perp p_1$ and $p_i\,\underset{\top}{\sim}\,q_i$. Then $p\,\underset{\top}{\sim}\,q$.*

Proof. Let s_i be a partial symmetry such that $s_i^2=p_i\vee q_i$, and $\{s_ip_is_i\}=q_i$, $i=1,2$. Since $p_1\vee q_1\perp p_2\vee q_2$, $s=s_1+s_2$ is a partial symmetry such that $s^2=p\vee q$. If we let $M=\bigoplus_{1\leq i\leq j\leq 3}M_{ij}$ be the Peirce decomposition of M corresponding to $p_1\vee q_1$, $p_2\vee q_2$, $1-p_1\vee q_1-p_2\vee q_2$, then s_i, p_i, $q_i\in M_{ii}$, $i=1,2$. Hence we have

$$\begin{aligned}\{sps\}&=\{s_1p_1s_1\}+\{s_1p_2s_1\}+\{s_2p_1s_2\}+\{s_2p_2s_2\}+2\{s_1p_1s_2\}+2\{s_1p_2s_2\}\\&=q_1+q_2=q,\end{aligned}$$

since $\{s_ip_is_i\}=q_i$, $i=1,2$, and the rest of the terms are zero.

5.2.11. Theorem. *Let M be a JBW algebra with projection lattice P. Let $p,q\in P$. Then we can write $p=p_1+p_2$, $q=q_1+q_2$ with p_i, $q_i\in P$, $i=1,2$, such that $p_1\,\underset{\top}{\sim}\,q_1$ and $c(p_2)\perp c(q_2)$.*

Proof. We first assume $p\perp q$. Let by Zorn's lemma $\{(p_\alpha,q_\alpha)\}_{\alpha\in J}$ be a maximal family of pairs of projections in M satisfying $p_\alpha\leqslant p$, $q_\alpha\leqslant q$, $p_\alpha\,\underset{\top}{\sim}\,q_\alpha$, all p_α are pairwise orthogonal, and all q_α are pairwise orthogonal. Let

$$p_1=\sum_{\alpha\in J}p_\alpha,\qquad q_1=\sum_{\alpha\in J}q_\alpha,\qquad p_2=p-p_1,\qquad q_2=q-q_1.$$

By 5.2.9 $p_1\,\underset{\top}{\sim}\,q_1$. By 5.2.7 and maximality of $\{(p_\alpha,q_\alpha)\}_{\alpha\in J}$ no nonzero subprojection of p_2 is equivalent to a subprojection of q_2. Since by 5.2.3(i) any two non-orthogonal projections contain nonzero equivalent subprojections it follows that for all $\beta\in\operatorname{Int}M$ (see 5.1.4), $\beta(p_2)\perp q_2$. Hence by 5.1.4 $c(p_2)=\bigvee_{\beta\in\operatorname{Int}M}\beta(p_2)\perp q_2$, and so $c(p_2)\perp c(q_2)$.

In the general case we apply the above to $p\wedge q^\perp$ and $q\wedge p^\perp$. Then we have

$$p\wedge q^\perp=p_{11}+p_2,\qquad q\wedge p^\perp=q_{11}+q_2,$$

where $p_{11} \underset{1}{\sim} q_{11}$ and $c(p_2) \perp c(q_2)$. Let $p_{12} = p - p \wedge q^{\perp}$, $q_{12} = q - q \wedge p^{\perp}$. By 5.2.3 $p_{12} \underset{1}{\sim} q_{12}$, so if we let $p_1 = p_{11} + p_{12}$, $q_1 = q_{11} + q_{12}$ an application of 5.2.10 shows that $p_1 \underset{1}{\sim} q_1$.

5.2.12. If $p, q \in P$ we write $p \underset{n}{\lesssim} q$ if there exists $q_0 \in P$, $q_0 \leqslant q$, *such that* $p \underset{n}{\sim} q_0$.

5.2.13. The comparison theorem. *Let M be a* JBW *algebra with projection lattice P. If p,q ∈ P there exists a central projection e in M such that*

$$ep \underset{1}{\lesssim} eq, \qquad e^{\perp}q \underset{1}{\lesssim} e^{\perp}p.$$

Proof. Let p_i and q_i be as in 5.2.11. Let $e = c(q_2)$. Then we have $ep = ep_1 \underset{1}{\sim} eq_1 \leqslant eq$, and $e^{\perp}q = e^{\perp}q_1 \underset{1}{\sim} e^{\perp}p_1 \leqslant e^{\perp}p$.

5.2.14. The halving lemma. *Let M be a* JBW *algebra with projection lattice P. Suppose M has no direct summand of type* I (cf. 5.1.5). *Then there is $p \in P$ with $p \underset{1}{\sim} p^{\perp}$.*

Proof. Let by Zorn's lemma $\{(p_\alpha, q_\alpha)\}_{\alpha \in J}$ be a maximal family of pairs of projections in M such that all p_α, q_α are pairwise orthogonal and $p_\alpha \underset{1}{\sim} q_\alpha$. By 5.2.9 $p = \sum_{\alpha \in J} p_\alpha \underset{1}{\sim} \sum_{\alpha \in J} q_\alpha = q$. We assert that $p + q = 1$. If not $N = \{(p+q)^{\perp} M (p+q)^{\perp}\}$ is nonzero. Since M has no type I part, N is not Abelian. Let r be a noncentral projection in N. Then $c(r)c(r^{\perp}) \neq 0$, so there exist by 5.2.11 nonzero projections $p_1 \leqslant r$, $q_1 \leqslant r^{\perp}$ in N such that $p_1 \underset{1}{\sim} q_1$. Since p_1 and q_1 are orthogonal to $p+q$ this contradicts the maximality of $\{(p_\alpha, q_\alpha)\}_{\alpha \in J}$, proving that $p + q = 1$, as asserted.

5.2.15. Proposition. *Let M be a* JBW *algebra with projection lattice P. Suppose M has no direct summand of type* I. *Then there are $p_i \in P$, $i =$* 1, 2, 3, 4, *such that $p_1 + p_2 + p_3 + p_4 = 1$, and $p_i \underset{1}{\sim} p_j$ for all i, j.*

Proof. By the halving lemma there are $p \in P$ and a symmetry s in M such that $\{sps\} = p^{\perp}$. Since an Abelian projection in $\{pMp\}$ is Abelian in M, the JBW algebra $\{pMp\}$ has no direct summand of type I, hence the halving lemma shows the existence of $p_1, p_2 \in P$ with $p_1 \underset{1}{\sim} p_2$ and $p_1 + p_2 = p$. Let $p_3 = \{sp_1s\}$, $p_4 = \{sp_2s\}$. Then clearly $p_1 + p_2 + p_3 + p_4 = 1$ and $p_3 \underset{1}{\sim} p_1 \underset{1}{\sim} p_2 \underset{1}{\sim} p_4$. An application of 5.2.6 completes the proof.

5.2.16. If $\mathcal{M}$ is a von Neumann algebra with centre $\mathcal{Z}$ it is quite easy, using the commutant, to show that if p is a projection in $\mathcal{M}$ then the centre of $p\mathcal{M}p$ is $\mathcal{Z}p$. To prove the same result for JBW algebras we have to use 5.2.11.

5.2.17. Proposition. *Let M be a JBW algebra with centre Z. Let p be a projection in M. Then the centre of $\{pMp\}$ is Zp.*

Proof. Let Z' denote the centre of $\{pMp\}$. Then clearly $Z' \supset Zp$. Let e be a projection in Z', and let $f = p - e$. Then f is a projection in Z', and by 5.2.11 there exist projections e_i, f_i, $i = 1, 2$, in M such that $e = e_1 + e_2$, $f = f_1 + f_2$, and $e_1 \sim f_1$, $c(e_2) \perp c(f_2)$. We assert that $e_1 = f_1 = 0$. Indeed, let s be a symmetry in M such that $U_s e_1 = f_1$, and let $t = \{psp\}$. Since $e_1, f_1 \in \{pMp\}$ we have by 2.4.18 that $U_t e_1 = U_p U_s U_p e_1 = U_p U_s e_1 = U_p f_1 = f_1$, so that by the positivity of U_f (3.3.6),

$$f_1 = U_f f_1 = U_f U_t e_1 \leq U_f U_t e.$$

Since f is central in $\{pMp\}$, f operator commutes with t, hence $U_t U_f = U_f U_t$. Therefore $f_1 \leq U_t U_f e = 0$, since $f \perp e$, and our assertion follows.

It follows that $c(e) \perp c(f)$. Since $c(e), c(f) \leq c(p)$ and $e + f = p$, we have $c(e) + c(f) = c(p)$. But then $e + f = p = c(e)p + c(f)p$. Since $c(e)p \geq e$ and $c(f)p \geq f$ we therefore have $e = c(e)p \in Zp$. Since e was an arbitrary projection in Z', and a JBW algebra is linearly generated by its projections (4.2.3), $Z' \subset Zp$, proving the proposition.

5.3. JBW algebras of type I

5.3.1. Among von Neumann algebras those of type I are the only ones which are completely understood. We shall see later that JBW algebras which are not of type I look very much like von Neumann algebras, hence they will roughly be known modulo the theory of von Neumann algebras of types II and III. JBW algebras of type I may be quite different from von Neumann algebras, cf. $H_3(\mathbb{O})$ and spin factors. We shall, however, succeed in the classification of JBW algebras of type I, the beginning of which will be done presently.

5.3.2. Lemma. *Let M be a JBW algebra with projection lattice P. Let $p, q \in P$.*

(i) *If p is Abelian and $q \leq p$, then $q = c(q)p$.*
(ii) *If both p and q are Abelian and $c(p) = c(q)$ then $p \sim q$.*
(iii) *If M is of type I then q dominates an Abelian projection r with $c(r) = c(q)$.*

Proof. (i) If p is Abelian then $\{pMp\}$ equals its centre, which by 5.2.17 is Zp, where Z is the centre of M. Thus if $q \leq p$ there is a central projection f in M such that $q = fp$. Since $fq = q$, $c(q) \leq f$, so $q = c(q)q = c(q)fp = c(q)p$.

(ii) By the comparison theorem (5.2.13) there is a central projection e such that $ep \precsim_1 eq$ and $e^\perp q \precsim_1 e^\perp p$. If p and q are Abelian with the same central supports then, since equivalent projections have the same central support, by (i) $ep \sim_1 eq$ and $e^\perp p \sim_1 e^\perp q$. Hence $p \sim_1 q$.

(iii) If M is of type I then there is by 5.1.4 an Abelian projection p with $c(p)=1$. By the comparison theorem (5.2.13) we may replace p with an equivalent projection and assume $p \leqslant q$ or $p \geqslant q$. If $p \leqslant q$ we may set $r=p$, while if $p \geqslant q$ then q itself is Abelian, so we can use $r=q$.

5.3.3. Let M be a JBW algebra and n a cardinal number. We say M is of *type* I_n if there is a family $(p_\alpha)_{\alpha\in J}$ of Abelian projections such that $c(p_\alpha)=1$, $\sum_{\alpha\in J} p_\alpha = 1$ and card $J=n$. We also say M is of type I_∞ if M is a direct sum of JBW algebras of type I_n with n infinite. It is clear that if M is of type I_n and e is a nonzero central projection in M then Me is of type I_n.

5.3.4. Lemma. *Let M be a* JBW *algebra of type* I_n, $n<\infty$. *Then n is the maximal cardinality of an orthogonal family of nonzero projections in M with the same central supports.*

Proof. We use induction on n. Since an algebra of type I_1 is associative, any projection is its own central support in this case, and the lemma is trivial. We assume it holds for JBW algebras of type I_{n-1}. Suppose that $q_1, \ldots, q_{n+1}$ are pairwise orthogonal projections in M with the same central supports, denoted by e. By 5.3.2(iii) we may assume each q_i is Abelian. Since Me is of type I_n there are Abelian projections $p_1, \ldots, p_n$ with central supports equal to e and sum e. By 5.3.2(ii) $q_{n+1} \sim_1 p_n$, say $p_n = \{sq_{n+1}s\}$ with s a symmetry in Me. Then $\{sq_is\}$, $i=1,\ldots,n$, are Abelian projections with central supports e dominated by $p_n^\perp$. This contradicts the inductive hypothesis and the fact that $\{p_n^\perp M p_n^\perp\}$ is of type I_{n-1}.

5.3.5. Theorem. *Each* JBW *algebra of type* I *has a unique decomposition*

$$M = M_1 \oplus M_2 \oplus \ldots \oplus M_\infty,$$

where each M_n is either 0 *or is a* JBW *algebra of type* I_n.

Proof. The uniqueness of the decomposition is immediate from 5.3.4. To show the existence let n be a cardinal number, and let $(e_\alpha)_{\alpha\in J}$ be a maximal family of pairwise orthogonal central projections such that $e_\alpha M$ is of type I_n. If $(e_\alpha)_{\alpha\in J}$ is empty let $M_n=0$; otherwise let $f_n=\sum_{\alpha\in J} e_\alpha$ and $M_n = f_n M$. Then clearly M_n is of type I_n, and by maximality of $(e_\alpha)_{\alpha\in J}$, $(1-f_n)M$ has no direct summand of type I_n. Let $f = f_\infty + \sum_{n\in\mathbb{N}} f_n$. We assert that $f=1$. If not, $1-f\neq 0$, and by considering $(1-f)M$ we may assume M has no direct summand of type I_n for any $n\in\mathbb{N}\cup\{\infty\}$. Let $(p_\beta)_{\beta\in I}$ be a maximal orthogonal family of Abelian projections with central supports 1, and let $p=1-\sum_{\beta\in I}p_\beta$. By 5.3.2(ii) $p_\beta\sim p_\gamma$ for all $\beta,\gamma\in I$. Let $n=\operatorname{card} I$. If $c(p)=1$ there exists by 5.3.2(iii) an Abelian projection with central support 1 dominated by p, contradicting the maximality of $(p_\beta)_{\beta\in I}$. Thus $c(p)<1$. But then $(1-c(p))M$ is of type I_n, contradicting our assumption on M. Thus $f=1$, and the proof is complete.

5.3.6. Recall from 4.6.1 that a JBW algebra is called a JBW factor if its centre consists of real multiples of the identity. The reader should note that many of the results obtained so far in this chapter take on a very simple form for JBW factors. In particular we have the following consequence of 5.3.5.

5.3.7. Corollary. *Let M be a* JBW *factor of type* I. *Then there exists $n\in\mathbb{N}\cup\{\infty\}$ such that M is of type I_n.*

5.3.8. Theorem. *Let M be a* JBW *factor of type I_n, $3\leq n<\infty$. Then M is isomorphic to one of $H_n(\mathbb{R})$, $H_n(\mathbb{C})$, $H_n(\mathbb{H})$ or $H_3(\mathbb{O})$ in the case $n=3$.*

Proof. By 5.3.2(i) each Abelian projection is minimal in M. Let $p_1,\ldots,p_n$ be pairwise orthogonal Abelian projections in M with sum 1. By 5.3.2(ii) and 5.2.8 they are strongly connected. Therefore by the coordinatization theorem (2.8.9) M is a Jordan matrix algebra $H_n(R)$with R a $*$ algebra. Since each p_i is minimal $R_{sa}\cong\mathbb{R}$. If we knew that R was finite-dimensional we could refer to 2.9.6. To avoid this approach we imitate the proof of 2.9.6. Let $\{e_{ij}: 1\leq i,j\leq n\}$ be the standard matrix units in $M_n(R)$. Suppose $0\neq x\in R$ and let $a=xe_{12}+x^*e_{21}$. Then $a\in M$, so $0\neq a^2=xx^*e_{11}+x^*xe_{22}$. Since $x^*x=\lambda 1$, $xx^*=\mu 1$, λ, $\mu\in\mathbb{R}$ not both zero, and since R is alternative by 2.7.6 $\lambda x=x(x^*x)=(xx^*)x=\mu x$, so $\lambda=\mu\neq 0$, and $x^*x\neq 0$. Thus the proof is completed by an application of 2.7.8.

5.3.9. Theorem. *Let M be a* JBW *algebra without direct summands of types I_2 and I_3. Then M is a* JW *algebra.*

Proof. By 5.1.5 and 5.3.5 M has a decomposition

$$M=M_{\mathrm{I}_1}\oplus M_{\mathrm{I}_4}\oplus\ldots\oplus M_{\mathrm{I}_\infty}\oplus M_{\mathrm{II}}\oplus M_{\mathrm{III}},$$

where each summand is of the type indicated. It suffices to show the theorem for each summand, and by 4.7.5 it suffices to show each summand is a JC algebra. Since M_{I_1} is associative this is immediate from 3.2.2 for that summand. We show that all the other summands contain at least four orthogonal projections with sum 1 which are exchanged by symmetries, hence are strongly connected (5.2.8), where we let 1 denote the identity in each summand. Indeed, in M_{I_n} there are n orthogonal Abelian projections with central supports 1 and sum 1. By 5.3.2(ii) they are all exchanged by symmetries in M_{I_n}. If $(p_\alpha)_{\alpha \in J}$ is an infinite orthogonal family of Abelian projections with central supports 1 and sum 1 we can write J as a disjoint union of four subsets $J_1, \ldots, J_4$ of equal cardinality. Let $q_i = \sum_{\alpha \in J_i} p_\alpha$. By 5.2.9 $q_i \curlyvee q_j$ for all i, j. Thus the assertion follows for M_{I_∞}. Finally in M_{II} and M_{III} the assertion follows from 5.2.15. By the coordinatization theorem (2.8.9) each summand is a Jordan matrix algebra of the form $H_n(R)$ with $n \geq 4$. An application of 3.7.2 completes the proof.

5.3.10. Theorem. *Let $\mathcal{M}$ be a von Neumann algebra and let M be a* JW *subalgebra of $\mathcal{M}_{sa}$ with no type* I_2 *part. Then M is reversible.*

Proof. Decomposing M as in 5.3.9 but with a I_3 part added, we see that it is sufficient to show that each summand is reversible. This is obvious for M_{I_1} since it is associative, while any other summand contains $n \geq 3$ pairwise orthogonal and strongly connected projections with sum 1 (see the proof of 5.3.9). The proof is completed by the 'special coordinatization theorem' 2.8.3, since $H_n(R)$ is clearly reversible in $M_n(R)$.

5.3.11. Let A be a JB algebra, N a JBW algebra and $\pi: A \to N$ a factor representation. We say π is of *type* I_n (resp. II, III) if $\pi(A)^-$ is of type I_n (resp. II, III).

5.3.12. Proposition. *Let M be a* JBW *algebra of type* I_n, $n < \infty$. *Then we have*:

(i) *Each factor representation of M is of type* I_n.

(ii) *If $n \geq 3$ and π a factor representation of M then $\pi(M)$ is a* JBW *algebra.*

Proof. (i) Let π be a factor representation of M. Let $p_1, \ldots, p_n$ be orthogonal Abelian projections with central supports 1 and sum 1 in M. By 5.3.2 $p_i \curlyvee p_j$ for all i, j, and hence $\pi(p_i) \curlyvee \pi(p_j)$ for all i, j. Since $\sum_{i=1}^n \pi(p_i) = 1$, and each $\pi(p_i)$ obviously is Abelian in $\pi(M)$ so in the weak closure $\pi(M)^-$, which is a JBW factor, $\pi(M)^-$ is of type I_n.

(ii) If $n \geq 3$ then by 5.3.8 $\pi(M)^-$ is finite-dimensional hence equal to $\pi(M)$. The proof is complete.

5.4. Comments

The main results in dimension theory in Section 5.2 were first proved for JW algebras by Topping [110]. They were then extended to JBW algebras in Alfsen *et al.* [19]. The decomposition results in Section 5.3 were also first proved for JW algebras by Topping [110], while 5.3.10 is due to Størmer [103].

6

Spin factors

6.1. Definition of spin factors

6.1.1. The canonical commutation and anticommutation relations play an important role in mathematical physics. For us the anticommutation relations will be of interest, because they give rise to a class of JC algebras of which we have only seen a glimpse, namely the spin factors encountered in Sections 2.9.7 and 2.9.8.

Let H be a real Hilbert space, and suppose $a: H \to \mathcal{A}$ is a linear map of H into a C^* algebra $\mathcal{A}$ satisfying the canonical anticommutation relations:

$$2a(f) \circ a(g) = a(f)a(g) + a(g)a(f) = 0,$$

$$2a(f) \circ a(g)^* = a(f)a(g)^* + a(g)^*a(f) = (f, g)1,$$

for all $f, g \in H$ (we use the notation f, g for elements in H since H is usually an L^2 space). If we let $b(f) = a(f) + a(f)^*$ then

$$b(f) \circ b(g) = (f, g)1, \qquad f, g \in H,$$

in particular $\|b(f)\| = \|f\|$, and b is an isometry of H into $\mathcal{A}_{sa}$ such that $b(H) + \mathbb{R}1$ is a JC algebra A. If $(f_n)_{n \in J}$ is an orthonormal basis for H then the set $\{1, b(f_n)\}_{n \in J}$ is a set of symmetries with the properties $b(f_n) \circ b(f_m) = \delta_{nm} 1$, which generate A linearly. The JC algebra A and the set $\{b(f_n)\}_{n \in J}$ are what we shall call a spin factor and a spin system respectively. Instead of pursuing the study of the anticommutation relations we shall rather study spin factors from the axiomatic point of view, and we refer the reader to Bratteli and Robinson [3, Ch. V] for the theory of the canonical anticommutation relations.

6.1.2. Let B be a real unital Jordan algebra. A *spin system* in B is a collection $\mathcal{P}$ of at least two symmetries different from ± 1 such that $s \circ t = 0$ whenever $s \neq t$ in $\mathcal{P}$. Given a spin system $\mathcal{P}$, let H_0 denote the linear span of $\mathcal{P}$ in B. Then any two elements a, b of H_0 can be written $a=\sum_{i=1}^n \alpha_i s_i$, $b=\sum_{i=1}^n \beta_i s_i$, where $s_1, \ldots, s_n$ are distinct symmetries in $\mathcal{P}$. From this we get

$$a \circ b = \left(\sum_{i=1}^{n} \alpha_i \beta_i\right) 1,$$

from which it follows that H_0 is a pre-Hilbert space with inner product defined by

$$\langle a, b\rangle 1 = a \circ b.$$

It is clear that $H_0 + \mathbb{R}1$ is a subalgebra of B. We shall be interested in the algebra obtained from this by completing H_0.

6.1.3. Lemma. *Let H be a real Hilbert space of dimension at least 2. Let $A = H \oplus \mathbb{R}1$ have the norm $\|a+\lambda 1\| = \|a\| + |\lambda|$, $a \in H$, $\lambda \in \mathbb{R}$. Define a product in A by*

$$(a+\lambda 1)\circ(b+\mu 1) = (\mu a + \lambda b) + (\langle a, b\rangle + \lambda\mu)1, \tag{6.1}$$

where $a, b \in H$, λ, $\mu \in \mathbb{R}$. Then A is a JB *algebra.*

Proof. To check the Jordan axiom $x\circ(y\circ x^2) = (x\circ y)\circ x^2$, note that x, y are contained in a subalgebra $H_2 \oplus \mathbb{R}1$ of A, where H_2 is a two-dimensional subspace of H. However, as was noticed in 2.9.8, the symmetric matrix algebra $H_2(\mathbb{R})$ is a spin factor; indeed, a basis for $H_2(\mathbb{R})$ is

$$1 = \begin{pmatrix} 1 & 0 \\ 0 & 1 \end{pmatrix}, \qquad \sigma_1 = \begin{pmatrix} 1 & 0 \\ 0 & -1 \end{pmatrix}, \qquad \sigma_2 = \begin{pmatrix} 0 & 1 \\ 1 & 0 \end{pmatrix},$$

where $\sigma_1^2 = \sigma_2^2 = 1$, $\sigma_1 \circ \sigma_2 = 0$. Thus, mapping an orthonormal basis of H_2 to σ_1, σ_2 we obtain an isomorphism of $H_2 \oplus \mathbb{R}1$ onto $H_2(\mathbb{R})$. Since $H_2(\mathbb{R})$ is a Jordan algebra, the Jordan axiom $x\circ(y\circ x^2) = (x\circ y)\circ x^2$ holds in $H_2 \oplus \mathbb{R}1$, and therefore in A. Pick an arbitrary element of $H_2(\mathbb{R})$, say $x = \lambda 1 + \mu_1\sigma_1 + \mu_2\sigma_2$. Then $(\mu_1^2 + \mu_2^2)^{-1/2}(\mu_1\sigma_1 + \mu_2\sigma_2)$ is a symmetry in $H_2(\mathbb{R})$ with both $+1$ and -1 in its spectrum, so by spectral theory

$$\|x\| = (\mu_1^2 + \mu_2^2)^{1/2} + |\lambda|.$$

Therefore, the above isomorphism of $H_2 \oplus \mathbb{R}1$ onto $H_2(\mathbb{R})$ is an isometry, and the JB algebra axioms $\|a\circ b\| \leq 1$, $\|a^2\| = \|a\|^2$, $\|a^2 + b^2\| \geq \|a^2\|$ follow because $H_2(\mathbb{R})$ is a JB algebra. This completes the proof.

Of course, all of the defining properties of a JB algebra could have been verified by direct calculation. An alternative approach would be to use the characterization of JB algebras as order unit spaces (3.1.6); see 6.1.6 below.

6.1.4. A unital JB algebra generated as a JB algebra by a spin system will be called a *spin factor.*

6.1.5. Proposition. *For each cardinal number $n \geqslant 2$ there is, up to isomorphism, a unique spin factor generated by a spin system of cardinality n.*

Proof. Given $n \geqslant 2$, the construction of 6.1.3 with $\dim H = n$ yields a spin factor generated by a spin system of cardinality n. Since H may be characterized as a subspace of A as the set of $a \in A$ for which $a \notin \mathbb{R}1$ while $a^2 \in \mathbb{R}1$, H, and therefore n, is uniquely determined by A.

Let A be a spin factor generated by a spin system $\mathcal{P}$ of cardinality n. By 6.1.2 the linear span H_0 of $\mathcal{P}$ is a pre-Hilbert space with inner product $\langle\ ,\ \rangle$ defined by $a \circ b = \langle a, b\rangle 1$. Note that if $\sum_{s \in \mathcal{P}} \alpha_s s \in H_0$, $\alpha_s \in \mathbb{R}$, then $\|\sum_s \alpha_s s\|^2 = \sum_s \alpha_s^2$, so the Hilbert space norm on H_0 coincides with the norm inherited from A. Thus if H denotes the Hilbert space completion of H_0 then $H \oplus \mathbb{R}1$ is a JB subalgebra of A, isomorphic to the spin factor defined in 6.1.3. Since $\mathcal{P}$ generates A, $A = H \oplus \mathbb{R}1$, and the proof is complete.

6.1.6. We call a symmetry in a JB algebra *nontrivial* if it is not ± 1, i.e. if its spectrum equals $\{-1, 1\}$. It is trivial to check that the nontrivial symmetries in a spin factor $A = H \oplus \mathbb{R}1$ are the unit vectors in H. In particular, if $a \in H$ then $\mathrm{Sp}(a) = \{-\|a\|, \|a\|\}$, and if $\lambda \in \mathbb{R}$, $a \in H$ then $\mathrm{Sp}(a + \lambda 1) = \{\lambda - \|a\|, \lambda + \|a\|\}$. Therefore

$$A^+ = \{a + \lambda 1 \colon a \in H, \lambda \in \mathbb{R}, \lambda \geqslant \|a\|\}.$$

This may be visualized as a circular cone.

6.1.7. Proposition. *Let A be a spin factor, and let H be the Hilbert space spanned by the nontrivial symmetries of A, so that $A = H \oplus \mathbb{R}1$. Then we have:*

(i) *There is a unique state τ on A annihilating H.*
(ii) *$\tau(x) = \tau(\{sxs\})$ for all symmetries $s \in A$ and all $x \in A$.*
(iii) *The uniform and strong topologies coincide on A.*
(iv) *A is its own second dual,* i.e. *$A = A^{**}$.*
(v) *Every homomorphism of A into a* JBW *algebra M is either 0 or an isomorphism onto a* JBW *subalgebra of M.*

Proof. Define τ by $\tau(a + \lambda 1) = \lambda$ for $a \in H$, $\lambda \in \mathbb{R}$. Then τ is a state by 6.1.6. Since uniqueness is clear (i) follows.

If $s \in A$ is a nontrivial symmetry $s \in H$, so if $a \in H$ then $U_s a = 2s \circ (s \circ a) - s^2 \circ a = 2\langle s, a\rangle s - a \in H$. Thus U_s maps H into itself so $\tau = \tau U_s$ by uniqueness of τ, and (ii) follows.

One of the (semi-)norms that define the strong topology on A is the norm $\| \ \|_2$ defined by τ,

$$\|a+\lambda 1\|_2^2=\tau((a+\lambda 1)^2)=\|a\|^2+|\lambda|^2, \qquad a\in H, \lambda\in\mathbb{R}.$$

This norm is clearly equivalent to the usual norm on A, indeed it satisfies the inequality

$$2^{-1/2}\| \ \|\leq\| \ \|_2\leq\| \ \|.$$

Thus the strong topology on A is at least as strong as the uniform topology, so they are equal, and (iii) is proved. The above inequality shows in particular that A is isomorphic to a Hilbert space, proving (iv).

Let M be a JBW algebra and let $\phi: A\to M$ be a nonzero homomorphism. Since $A=A^{**}$, ϕ is normal by 4.5.7. By 4.5.11 $\phi(A)$ is a JBW subalgebra of M. To show that ϕ is injective, assume $\phi(a+\lambda 1)=0$, $a\in H$, $\lambda\in\mathbb{R}$. Pick a nonzero vector $b\in H$ orthogonal to a. Then

$$0=\phi(a+\lambda 1)\circ\phi(b)=\phi((a+\lambda 1)\circ b)=\lambda\phi(b),$$

so if $\lambda\neq 0$, then $\phi(b)=0$, and so $\|b\|^2\phi(1)=\phi(b^2)=0$. But then $\phi(1)=0$, so $\phi=0$. Similarly, if $\lambda=0$ then $\|a\|^2\phi(1)=0$, so if $\phi\neq 0$ then $a=0$. This proves that ϕ is injective.

6.1.8. Theorem. *Let M be a* JBW *algebra. Then M is a* JBW *factor of type* I_2 *if and only if M is a spin factor.*

Proof. Suppose M is a spin factor, $M=H\oplus\mathbb{R}1$. If e is a central projection in M then the map T_e is a normal homomorphism of M into a JBW algebra (2.5.6), hence it is 0 or injective by 6.1.7. Thus M is a JBW factor. If M is not of type I_2 there are three nonzero pairwise orthogonal projections p_1, p_2, p_3 in M, cf. 5.2.15 and 5.3.3. But then the element p_1+2p_2 has spectrum equal to $\{0, 1, 2\}$, contradicting the conclusion in 6.1.6 that the spectrum of an element in M has cardinality at most 2. Thus M is of type I_2.

Conversely let M be a JBW factor of type I_2. Let H denote the linear span of all symmetries in M different from ± 1. Then $M=H+\mathbb{R}1$. Indeed, since M is of type I_2, if $a\in M$ then by the spectral theorem (3.2.4) there are minimal projections p, q in M with sum 1 and $\alpha, \beta\in\mathbb{R}$ such that

$$a=\alpha p+\beta q=\tfrac{1}{2}(\alpha+\beta)1+\tfrac{1}{2}(\alpha-\beta)(p-q), \tag{6.2}$$

hence $a\in H+\mathbb{R}1$, as asserted.

Let $s, t\in H$ be symmetries different from ± 1. Then $s\circ t\in\mathbb{R}1$. Indeed, let $p, q\in M$ be minimal projections such that $s=p-q$. Let $M=M_1\oplus M_{1/2}\oplus M_0$ be the Peirce decomposition of M corresponding to p. Then $t=\alpha p+r+\beta q$ with $r\in M_{1/2}$ and $\alpha, \beta\in\mathbb{R}$. Since $p\circ r=q\circ r=\frac{1}{2}r$ we have $1=t^2=\alpha^2p+\beta^2q+r^2+(\alpha+\beta)r$. By 2.6.3 $r^2\in M_1\oplus M_0$ so that $(\alpha+\beta)r=0$. If $r=0$ then $\alpha^2=\beta^2=1$, so that $\alpha=-\beta$ since $t\neq\pm1$. If $r\neq0$ then $\alpha+\beta=0$. Thus in either case $t=\alpha(p-q)+r$, so that $s\circ t=(p-q)\circ(\alpha(p-q)+r)=\alpha 1\in\mathbb{R}1$, proving the assertion.

If $a=\sum\alpha_is_i$, $b=\sum\beta_jt_j\in H$, α_i, $\beta_j\in\mathbb{R}$, s_i, t_j symmetries different from ±1, then it follows from the above that $a\circ b\in\mathbb{R}1$. Since $a^2\geqslant0$, $a^2=\alpha1$ with $\alpha\geqslant0$. Therefore $\langle a, b\rangle 1=a\circ b$ defines an inner product $\langle\ ,\ \rangle$ on H making H into a real pre-Hilbert space such that the Jordan product in M is given by (6.1). We shall complete the proof by first showing that M is the direct sum of $\mathbb{R}1$ and H, and then that H is complete.

Suppose $1=\sum_{i=1}^n\alpha_is_i\in H$, $\alpha_i\in\mathbb{R}$, s_i symmetries different from ±1. Then for $1\leqslant j\leqslant n$ we have

$$s_j=s_j\circ1=\sum_{i=1}^m\alpha_is_j\circ s_i\in\mathbb{R}1,$$

contrary to assumption. Thus the sum $H+\mathbb{R}1$ is direct.

Finally let τ be the linear functional on M defined by $\tau(1)=1$ and τ vanishes on H. By (6.2) τ is positive, hence a state. Therefore by the definition of the order norm in an order unit space (1.2.1) and the fact that this norm coincides with the norm $\|\ \ \|$ on M (3.3.10), τ is continuous. Thus its null space H is closed, so H is a Hilbert space. Since M is nonassociative the dimension of H is at least 2, hence M is a spin factor.

6.2. Representations of spin factors

6.2.1. We shall show that each spin factor has a faithful representation as a JW factor, i.e. a JW algebra which is also a JBW factor. By 6.1.6 it suffices to construct for each cardinal number α a spin system of cardinality α consisting of symmetries in a C^* algebra. We shall first do this for finite and countably infinite spin systems.

Let 1_n denote the identity in $M_n(\mathbb{C})$. We define the tensor product $M_n(\mathbb{C})\otimes M_m(\mathbb{C})$ to be $M_{nm}(\mathbb{C})$ identified with $M_m(M_n(\mathbb{C}))$, in which the imbeddings of $M_n(\mathbb{C})$ and $M_m(\mathbb{C})$ are as follows: if $a\in M_n(\mathbb{C})$ then

$$a\otimes1_m=\begin{pmatrix}a & & 0\\ & \ddots & \\ 0 & & a\end{pmatrix}\in M_m(M_n(\mathbb{C})),$$

is the $m\times m$ matrix with a repeated down the diagonal and zero otherwise. If $b=(b_{ij})\in M_m(\mathbb{C})$ then

$$1_n\otimes b=(b_{ij}1_n)\in M_m(M_n(\mathbb{C})),$$

is the $m\times m$ matrix with entries $b_{ij}1_n$. The tensor product $a\otimes b$ is then the matrix

$$a\otimes b=(a\otimes 1_m)(1_n\otimes b)=(b_{ij}a)\in M_m(M_n(\mathbb{C})).$$

With the imbedding $M_n(\mathbb{C})\to M_{nm}(\mathbb{C})$ by $a\to a\otimes 1_m$ we write $M_n(\mathbb{C})\hookrightarrow M_{nm}(\mathbb{C})$ and consider $M_n(\mathbb{C})$ as contained in $M_{nm}(\mathbb{C})$. If we apply this to tensor products of $M_2(\mathbb{C})$ with itself we obtain an inductive system

$$M_2(\mathbb{C})\to M_{2^2}(\mathbb{C})\to M_{2^3}(\mathbb{C})\to\ldots.$$

The inductive limit $\mathcal{A}_0$ of this inductive system is a $*$ algebra. By uniqueness of the C^* norm in each $M_{2^n}(\mathbb{C})$ we obtain a C^* norm $\|\ \|$ on $\mathcal{A}_0$ by taking $\|x\|$ to be the norm of y in $M_{2^n}(\mathbb{C})$ whenever x has a representative y in $M_{2^n}(\mathbb{C})$. The completion of $\mathcal{A}_0$ in this norm is a C^* algebra, usually called the CAR *algebra* (CAR being shorthand for canonical anticommutation relations). $\mathcal{A}$ is clearly a separable C^* algebra. It is simple, for if π is a $*$ representation of $\mathcal{A}$ then, since each $M_{2^n}(\mathbb{C})$ is simple, π restricted to each $M_{2^n}(\mathbb{C})$ is an isometry. Thus π is an isometry on $\mathcal{A}_0$ and therefore on $\mathcal{A}$.

For each $n\in\mathbb{N}$ and for the countably infinite cardinal $\aleph_0$ we shall construct a spin factor which generates $M_{2^n}(\mathbb{C})$, $M_{2^n}(\mathbb{C})\oplus M_{2^n}(\mathbb{C})$, or $\mathcal{A}$ respectively. Let

$$\sigma_1=\begin{pmatrix}1 & 0\\ 0 & -1\end{pmatrix},\qquad \sigma_2=\begin{pmatrix}0 & 1\\ 1 & 0\end{pmatrix},\qquad \sigma_3=\begin{pmatrix}0 & i\\ -i & 0\end{pmatrix}$$

be the Pauli spin matrices in $M_2(\mathbb{C})$. (This is why we denote our spin systems by $\mathcal{P}$.) Denote by σ_3^n the n-fold tensor product $\sigma_3\otimes\ldots\otimes\sigma_3$ of σ_3 with itself n times in $M_{2^n}(\mathbb{C})$. Define

$$\begin{aligned}
s_1&=\sigma_1\\
s_2&=\sigma_2\\
s_3&=\sigma_3\otimes\sigma_1\\
s_4&=\sigma_3\otimes\sigma_2\\
&\ \vdots\\
s_{2n+1}&=\sigma_3^n\otimes\sigma_1\\
s_{2n+2}&=\sigma_3^n\otimes\sigma_2.
\end{aligned}$$

With the imbeddings $M_{2^n}(\mathbb{C})\hookrightarrow M_{2^{n+1}}(\mathbb{C})$ defined previously we have $s_k\in M_{2^n}(\mathbb{C})$ if $k\leqslant 2n$. Since $\sigma_i\circ\sigma_j=\delta_{ij}1_2$ it follows that $\mathcal{P}_k=\{s_1,\ldots,s_k\}$ is a spin system for each $k\in\mathbb{N}$. The linear span V_k of 1 and $\mathcal{P}_k$ is then a $(k+1)$-dimensional spin factor contained in $M_{2^n}(\mathbb{C})$ if $k\leqslant 2n$. Since the

C^* algebra generated by σ_1 is the diagonal matrices in $M_2(\mathbb{C})$, and the C^* algebra generated by σ_1 and σ_2 is $M_2(\mathbb{C})$, it follows that the C^* algebra $C^*(V_k)$ generated by V_k equals $M_{2^{n-1}}(\mathbb{C})\oplus M_{2^{n-1}}(\mathbb{C})$ if $k=2n-1$ and equals $M_{2^n}(\mathbb{C})$ if $k=2n$.

Since each s_k belongs to the CAR algebra $\mathcal{A}$ the spin system $\{s_1, s_2, \ldots\}$ generates an infinite-dimensional spin factor V_∞ inside $\mathcal{A}$. Since $s_1, \ldots, s_{2n}$ generate $M_{2^n}(\mathbb{C})$ as a C^* algebra it follows that $C^*(V_\infty)=\mathcal{A}$.

We summarize our results in the following theorem.

6.2.2. Theorem. *For each $k\in\{2, 3, \ldots, \infty\}$ there exists a spin factor V_k of dimension $k+1$ contained in $M_{2^n}(\mathbb{C})$ if $k\in\{2n-1, 2n\}$ and contained in the* CAR *algebra $\mathcal{A}$ if $k=\infty$. The C^* algebra $C^*(V_k)$ generated by V_k is described as follows:*

$$C^*(V_k)=\begin{cases} M_{2^{n-1}}(\mathbb{C})\oplus M_{2^{n-1}}(\mathbb{C}) & \text{if } k=2n-1,\\ M_{2^n}(\mathbb{C}) & \text{if } k=2n,\\ \mathcal{A} & \text{if } k=\infty.\end{cases}$$

6.2.3. Theorem. *Let A be a* JB *algebra of real dimension at least* 3. *Then A is a spin factor if and only if A is a* JW *factor of type* I_2.

Proof. By 6.1.8 it remains to show the necessity, so assume A is a spin factor. We show that A is a JC algebra, because then an application of 4.7.5 completes the proof. If the dimension of A is finite or countably infinite this follows from 6.2.2. Let A be obtained from a spin system $\mathcal{P}$ of arbitrary infinite cardinality. Let $\mathcal{F}$ denote the family of all finite subsets of $\mathcal{P}$ ordered by inclusion. For each $F\in\mathcal{F}$ denote by $A(F)$ the spin factor obtained from the spin system F. By 6.2.2 $A(F)$ is a JC algebra generating a C^* algebra $C^*(F)$, and there is a natural imbedding of $C^*(F)$ into $C^*(F')$ carrying $A(F)$ into $A(F')$ whenever $F\subset F'$. We thus have an inductive system of C^* algebras $\{C^*(F)\}_{F\in\mathcal{F}}$ whose inductive limit is as for the $M_{2^n}(\mathbb{C})$ a normed $*$ algebra whose completion is a C^* algebra $C^*(\mathcal{F})$. Since $C^*(\mathcal{F})$ contains $A(F)$ for all $F\in\mathcal{F}$ we obtain a representation of A as a JC algebra, completing the proof.

6.2.4. By 5.3.10 every JW factor which is not of type I_2 is reversible, hence the only possibility for nonreversible JW factors is spin factors. The next result is a solution to this problem.

6.2.5. Theorem. *Let A be a spin factor. If A has a representation as a reversible* JC *algebra then A is isomorphic to one of the spin factors V_2, V_3 or V_5. Conversely V_2 and V_3 are reversible in every representation in a C^* algebra, while V_5 has both reversible and nonreversible representations.*

Proof. Suppose A is a reversible spin factor inside a C^* algebra. Write $A = H \oplus \mathbb{R}1$ as in 6.1.3 and assume first $\dim H \geqslant 4$. Choose a set $s_1, \ldots, s_4$ of symmetries in H such that $\frac{1}{2}(s_i s_j + s_j s_i) = s_i \circ s_j = 0$, and let $s_5 = s_1 s_2 s_3 s_4$. Since $s_i s_j = -s_j s_i$ for $i \neq j$, $s_5 = \frac{1}{2}(s_1 s_2 s_3 s_4 + s_4 s_3 s_2 s_1)$, so $s_5 \in A$, since A is reversible. Furthermore $s_5 \circ s_i = 0$ if $i \leqslant 4$, as is easy to check. Thus $\{s_1, \ldots, s_5\}$ is a spin system in H, which is impossible if $\dim H = 4$, i.e. if $A = V_4$. Therefore V_4 is never reversible.

If $\dim H \geqslant 6$ choose a new symmetry s_6 in H such that $s_6 \circ s_i = 0$, $i \leqslant 5$. Then we have

$$s_5 s_6 = -s_6 s_5 = -s_6 s_1 s_2 s_3 s_4 = -s_1 s_2 s_3 s_4 s_6 = -s_5 s_6,$$

which is impossible. Thus the only possibilities for A with reversible representations are V_2, V_3 and V_5.

Suppose V_3 is contained in a C^* algebra. Let $\{s_1, s_2, s_3\}$ be a spin system of symmetries defining V_3. In order to show V_3 is reversible it suffices by multilinearity to prove that $x = a_1 \ldots a_n + a_n \ldots a_1 \in V_3$ whenever a_i belongs to the basis $\{1, s_1, s_2, s_3\}$ for V_3. Using the cancellations $s_i^2 = 1$, $s_i s_j = -s_j s_i$, $i \neq j$, we may permute the a_i's (possibly reversing a sign in the expression for x) and cancel terms, until we find $x = b_1 \ldots b_m + b_m \ldots b_1$, $b_i \in A$, where $m \leqslant 3$. If $m = 2$, $x = 2 b_1 \circ b_2 \in A$, and if $m = 3$, $x = 2\{b_1 b_2 b_3\} \in A$. Thus $x \in V_3$, and V_3 is reversible in every representation. Similarly V_2 is reversible in every representation.

V_5 can be both reversible and nonreversible in concrete representations. Indeed, $H_2(\mathbb{H})$ having real dimension 6 is a reversible representation of V_5. From the first paragraph of the proof $H_2(\mathbb{H})$ is obtained from a spin system $\{s_1, \ldots, s_5\}$ where $s_5 = s_1 s_2 s_3 s_4$. Since two spin factors of the same dimension are isomorphic (6.1.5) there is a representation π of $H_2(\mathbb{H})$ such that $\pi(s_i) = s_i$ if $i = 1, \ldots, 4$, and $\pi(s_5) = -s_5$. Then $\psi = \iota \oplus \pi$, with ι the identity map, is a representation of V_5 in $M_8(\mathbb{C})$ satisfying

$$\psi(a) = \begin{pmatrix} a & 0 \\ 0 & \pi(a) \end{pmatrix}.$$

We have

$$\begin{aligned} &\tfrac{1}{2}[\psi(s_1) \ldots \psi(s_5) + \psi(s_5) \ldots \psi(s_1)] \\ &= \tfrac{1}{2}\begin{pmatrix} s_1 \ldots s_5 + s_5 \ldots s_1 & 0 \\ 0 & -(s_1 \ldots s_5 + s_5 \ldots s_1) \end{pmatrix} \\ &= \begin{pmatrix} 1 & 0 \\ 0 & -1 \end{pmatrix} \notin \psi(V_5), \end{aligned}$$

proving that $\psi(V_5)$ is not reversible.

6.3. JBW algebras of type I_2

6.3.1. By 6.1.8 each JBW factor of type I_2 is a spin factor and so by 6.2.3 in particular a JW algebra. We shall in the present section characterize general JBW algebras of type I_2 and obtain direct global generalizations of our previous results.

6.3.2. Lemma. *Let M be a JBW algebra of type I_2 and p a projection in M. Then p is Abelian if and only if $c(p^\perp)=1$.*

Proof. Since $c(p^\perp)\geqslant p^\perp$, $c(p^\perp)^\perp\leqslant p$. Thus if p is Abelian then by 5.3.2 so is $c(p^\perp)^\perp$. But M is of type I_2, so no nonzero central projection can be Abelian. Therefore $c(p^\perp)^\perp=0$. Conversely, if $c(p^\perp)=1$ and p is not Abelian we can by 5.3.2 choose nonzero projections $q, r\leqslant p$ with $q\perp r$ and $q\sim r$. Then $p^\perp, q, r$ are orthogonal projections with $c(p^\perp)c(q)c(r)\neq 0$, which is impossible by 5.3.4.

6.3.3. If M is a JBW algebra of type I_2 with centre Z we may by 3.2.2 assume $Z=C(X)$ with X a compact Hausdorff space. Then an element a in M can be considered as a function on X with values in JBW factors of type I_2. (We shall make this more precise in 6.3.13 below, but it helps the intuition to look at elements in M this way.) Therefore, for each $t\in X$, $a(t)=x(t)p(t)+y(t)q(t)$, where $x(t), y(t)\in\mathbb{R}$, and $p(t)$ and $q(t)$ are Abelian projections in a I_2 factor with sum 1. We should thus expect that there exist $x, y\in Z$ and Abelian projections p and q in M with sum 1 such that $a=xp+yq$. The next result is a formal proof of this fact.

6.3.4. Lemma. *Let M be a JBW algebra of type I_2 with centre Z. If $a\in M$ there exist $x,y\in Z$ and a projection $p\in M$ with $c(p)=c(p^\perp)=1$ such that $a=xp+yp^\perp$. In particular, if s is the symmetry $s=2p-1$ then $a=z+ws$ with $z,w\in Z$.*

Proof. Define $z_1, z_2\in Z$ by

$$z_1=\max\{z\in Z: z\leqslant a\}, \qquad z_2=\min\{z\in Z: z\geqslant a\}. \tag{6.3}$$

Then $z_1\leqslant a\leqslant z_2$. By spectral theory (3.2.2) and the fact that Z is a JBW algebra we can find central projections $e_0, e_1, e_2, \ldots$ with sum 1 such that $e_0z_1=e_0z_2$, and real numbers $s_n<t_n$ for $n\geqslant 1$, such that $z_1e_n\leqslant s_ne_n\leqslant t_ne_n\leqslant e_nz_2$. As in 4.1.10 let $W(a)$ be the JBW algebra generated by a. By 4.1.11 we can identify $W(a)$ with $C(Y)$ for some compact Hausdorff space Y. Let $s_n<\lambda_n<t_n$ and $Y_n=\{y\in Y: a(y)\leqslant\lambda_n\}$, and let p_n be the projection in $W(a)$ corresponding to the characteristic function for Y_n.

We assert that

$$e_n \leqslant c(p_n)c(p_n^\perp). \tag{6.4}$$

Let $f = e_n c(p_n)^\perp$ and put $z = \lambda_n f + z_1 f^\perp$. Since $fp_n = 0$ we have $\lambda_n f = \lambda_n f p_n^\perp \leqslant f p_n^\perp a$ by definition of p_n, so that $z \leqslant a$. By (6.3) $z \leqslant z_1$, and in particular $\lambda_n f = zf \leqslant z_1 f$. But $\lambda_n f \geqslant s_n f \geqslant z_1 f$ since $f \leqslant e_n$, so we must have $\lambda_n f = s_n f$. Since $s_n < \lambda_n$ this implies $f = 0$, hence $e_n \leqslant c(p_n)$.

In order to show $e_n \leqslant c(p_n^\perp)$ we use a symmetric argument; just replace z_1 by z_2 and s_n by t_n in the above paragraph and use inequalities in the opposite direction.

By 4.3.5 $c(e_n p_n) = e_n c(p_n) = e_n$, and similarly $c(e_n p_n^\perp) = e_n$. Thus by 6.3.2 $e_n p_n$ and $e_n p_n^\perp$ are Abelian projections in $e_n M$. Since by 5.2.17 the centre of $\{p_n M p_n\}$ is Zp_n, we find $e_n\{p_n M p_n\} = \{e_n p_n M e_n p_n\} = e_n p_n Z$. Since $c(e_n p_n) = e_n$ the map $e_n Z \to e_n p_n Z$ is by 4.3.4 an isomorphism. Since $p_n \in W(a)$ it operator commutes with a, hence there is $x_n \in Z$ with $\|x_n\| \leqslant \|a\|$ such that $e_n(p_n \circ a) = e_n\{p_n a p_n\} = x_n e_n p_n$. Similarly there is $y_n \in Z$ with $\|y_n\| \leqslant \|a\|$ such that $e_n(p_n^\perp \circ a) = y_n e_n p_n^\perp$. Adding we get $e_n a = x_n e_n p_n + y_n e_n p_n^\perp$.

Let p_0 be any Abelian projection in M with $c(p_0) = 1$, and let $x_0 = y_0 = e_0 z_1$. Then $x_0 = y_0 = e_0 a$. Therefore, if $p = \sum_{n=0}^\infty e_n p_n$, $x = \sum_{n=0}^\infty x_n e_n$, $y = \sum_{n=0}^\infty y_n e_n$, then by (6.4) $c(p) = \sum c(e_n p_n) = \sum e_n = 1$ and similarly $c(p^\perp) = 1$. Furthermore we have

$$a = \sum e_n a = \sum (x_n e_n p_n + y_n e_n p_n^\perp) = xp + yp^\perp,$$

proving the first assertion of the lemma.

Finally, if $s = 2p - 1$, $z = \frac{1}{2}(x + y)$, $w = \frac{1}{2}(x - y)$, then s is a symmetry, $z, w \in Z$, and $a = z + ws$. The proof is complete.

6.3.5. If $s = 2p - 1$ is a symmetry then $s + 1 = 2p$ and $s - 1 = 2p^\perp$. We write $c(s+1)$ and $c(s-1)$ for the central supports of p and $p^\perp$ respectively. Thus the symmetry s of 6.3.4 has the property that $c(s \pm 1) = 1$.

6.3.6. Lemma. *Let M be a* JBW *algebra of type* I_2 *with centre Z. Let $N = \{ws\colon w \in Z,\ s$ a symmetry in M with $c(s \pm 1) = 1\}$. If s, t are symmetries in N then $s \circ t \in Z$.*

Proof. We have $s = p - p^\perp$ for an Abelian projection p with $c(p) = c(p^\perp) = 1$. Let $M = M_1 \oplus M_{1/2} \oplus M_0$ be the Peirce decomposition of M corresponding to p. Since p and $p^\perp$ are Abelian, by 5.2.17 $M_1 = \{pMp\} = Zp$ and $M_0 = \{p^\perp M p^\perp\} = Zp^\perp$. It follows that the Peirce decomposition of t is of the form

$$t = up + vp^\perp + r, \qquad u, v \in Z, \quad r \in M_{1/2}. \tag{6.5}$$

Then we have

$$1=t^2=u^2p+v^2p^\perp+r^2+2(up+vp^\perp)\circ r. \tag{6.6}$$

Hence the first three terms belong to $M_1\oplus M_0$, while the last term belongs to $M_{1/2}$ and thus is 0. Since $up\circ r=\frac{1}{2}ur$ and $vp^\perp\circ r=\frac{1}{2}vr$, the last term is $(u+v)r$, so that if $e\in Z$ denotes the support projection of $u+v$, we have $er=0$. If we multiply (6.6) by e we therefore obtain $e=eu^2p+ev^2p^\perp$, and so $eu^2p=ep$. Since $c(p)=1$ the map $Z\to Zp$ is an isomorphism (4.3.4), hence $eu^2=e$. Similarly $ev^2=e$. Therefore there is a central projection f such that $efu=efv$ and $ef^\perp u=-ef^\perp v$, and it follows from (6.5) that

$$eft=efup+efvp^\perp=efv\in Z.$$

Now $(eft)^2=ef$, so that eft is a central partial symmetry. Since $t\in N$, $c(t\pm 1)=1$, so that by 4.3.5 $c(ef(t\pm 1))=ef$. But this is impossible unless $ef=0$, since eft is a central partial symmetry. We therefore conclude that $ef=0$, and so $eu=-ev$.

By definition of e as the support projection of $u+v$ we have $(u+v)e^\perp=0$, so that $ue^\perp=-ve^\perp$. But then it follows that $u=-v$, and therefore

$$\begin{aligned}s\circ t&=p\circ t-p^\perp\circ t=up+p\circ r-vp^\perp-p^\perp\circ r\\&=u(p+p^\perp)+\tfrac{1}{2}r-\tfrac{1}{2}r=u\in Z.\end{aligned}$$

The proof is complete.

6.3.7. Proposition. *Let M be a JBW algebra of type I_2 with centre Z. Let N be as in 6.3.6. Then N is a vector subspace of M, and $M=Z\oplus N$.*

Proof. Let $z\in Z$ and s be a symmetry in N. Suppose $zs\in N\cap Z$. Then $zs=0$. Indeed, if $z\neq 0$ there is by the spectral theorem (3.2.4) a projection $e\in Z$ such that ez is invertible in eZ. But then $es\in Z$, which contradicts the assumption that $s\in N$, so that $c(s\pm 1)=1$. Thus $z=0$, and in particular $N\cap Z=\{0\}$.

Let next s_1 and s_2 be symmetries in N and $w_1, w_2\in Z$. By 6.3.4 there are $z, w_3\in Z$ and a symmetry $s_3\in N$ such that $w_1s_1+w_2s_2=z+w_3s_3$. If we multiply this equation with any symmetry $s\in N$ and use 6.3.6 we obtain $zs\in N\cap Z$, so $z=0$ by the preceding paragraph. But then $w_1s_1+w_2s_2=w_3s_3\in N$, and N is a vector subspace of M. Since by 6.3.4 $M=Z+N$, the proof is complete.

6.3.8. Let M be a JBW algebra of type I_2. A spin system $(s_\alpha)_{\alpha\in J}$ in M is called *locally maximal* if for every nonzero central projection e in M the family $(es_\alpha)_{\alpha\in J}$ is a maximal spin system in eM.

Note that since card $J \geq 2$ then $c(s_\alpha \pm 1) = 1$ for all $\alpha \in J$. Indeed if $s_\beta = p - p^\perp$ and $c(p^\perp) < 1$ (resp. $c(p) < 1$) then there is a nonzero central projection e such that $es_\beta = e$ (resp. $= -e$), hence $(es_\alpha)_{\alpha \in J}$ is not a spin system, a contradiction.

6.3.9. Lemma. *Let M be a JBW algebra of type I_2 with centre Z. Suppose $(s_1, \ldots, s_k)$, $k \in \mathbb{N}$, is a locally maximal spin system in M. Then every operator $a \in M$ can be written uniquely in the form*

$$a = z_0 + \sum_{i=1}^{k} z_i s_i, \qquad z_0, \ldots, z_k \in Z.$$

Proof. Let $a \in M$. By 6.3.7 there are unique $z_0 \in Z$ and $ws \in N$ with $w \in Z$, s a symmetry in N, such that $a = z_0 + ws$. By 6.3.6 $s \circ s_i \in Z$ for all s_i. Let $b = s - \sum_{i=1}^{k} (s \circ s_i) s_i$. By 6.3.7 $b \in N$, so $b = vt$ with $v \in Z$ and t a symmetry in N. Since $b \circ s_i = 0$ for all i, $0 = (vt) \circ s_i = v(t \circ s_i)$. Let e denote the support projection of v. Then $e \in Z$ and $e(t \circ s_i) = 0$ for all i. Thus $(et, es_1, \ldots, es_k)$ is a spin system in eM, contradicting the local maximality of $(s_1, \ldots, s_k)$. Therefore $e = 0$, so $v = 0$, and therefore $s = \sum_{i=1}^{k} (s \circ s_i) s_i$, proving that a can be written in the desired form.

To show uniqueness assume $a = 0$. Since by 6.3.7 z_0 is unique, $z_0 = 0$, and $\sum_{i=1}^{k} z_i s_i = 0$. But then we have $z_j = (\sum_{i=1}^{k} z_i s_i) \circ s_j = 0$.

6.3.10. Let M be a JBW algebra of type I_2. Let $k \in \{2, 3, \ldots\} \cup \{\infty\}$. We say M is of *type* $I_{2,k}$ if there is a locally maximal spin system $(s_\alpha)_{\alpha \in J}$ in M with card $J = k$ if $k < \infty$, and card J is infinite if $k = \infty$.

The number k is uniquely defined. This will follow from our next result. Note that by 4.6.4 and 5.3.12 each JBW algebra of type I_2 has a separating family of factor representations of type I_2. Since each JBW factor of type I_2 is a spin factor (6.1.8), and each spin factor is a JC algebra (6.2.3), it follows from 4.7.5 that each JBW algebra of type I_2 is a JW algebra.

6.3.11. Proposition. *Let M be a JBW algebra of type $I_{2,k}$, $k < \infty$. Then each factor representation of M is onto the spin factor V_k.*

Proof. Let π be a factor representation of M. Since by 5.3.12 $\pi(M)^-$ is of type I_2 it is a spin factor (6.1.8), hence by 6.1.7 $\pi(M) = \pi(M)^-$ since $\pi(M)$ is norm-closed by 3.4.3. Let $\{s_1, \ldots, s_k\}$ be a locally maximal spin system in M. Then the symmetries $\pi(s_1), \ldots, \pi(s_k)$ all anticommute, i.e. $\pi(s_i) \circ \pi(s_j) = 0$ if $i \neq j$, hence they form a spin system in $\pi(M)$. Since $\pi(Z) = \mathbb{R}1$, where Z denotes the centre of M, it follows by 6.3.9 that each element in $\pi(M)$ can be written in the form $\alpha_0+\sum_{i=1}^k \alpha_i\pi(s_i)$, which proves that $\pi(M)\cong V_k$.

6.3.12. If N is a JBW algebra and X a compact Hausdorff space then $C(X, N)$ denotes the set of continuous functions on X with values in N. If $C(X, N)$ is given pointwise addition and multiplication it is clearly a Jordan algebra. If the norm is the sup-norm it is a JB algebra. We then have the following variation of 6.3.11, which answers the comments in 6.3.3.

6.3.13. Proposition. *Let M be a JBW algebra of type $I_{2,k}$, $k<\infty$. Let Z denote the centre of M, and by 3.2.2 let X be a compact Hausdorff space such that $Z\simeq C(X)$. Then $M\simeq C(X, V_k)$.*

Proof. Let $(s_1,\ldots,s_k)$ be a locally maximal spin system in M. Let $(t_1,\ldots,t_k)$ be a spin system generating V_k. Then $(1, t_1,\ldots,t_k)$ is a basis for V_k, so if $a\in C(X, V_k)$ then $a(t)=z_0(t)1+\sum_{i=1}^k z_i(t)t_i$ for $t\in X$, where $z_0,\ldots,z_k$ are continuous real functions on X. Thus $z_0,\ldots,z_k\in C(X)\simeq Z$, and the map $a\to z_0+\sum_{i=1}^k z_is_i$ is by 6.3.9 an isomorphism of $C(X, V_k)$ onto M.

6.3.14. Theorem. *Any JBW algebra of type I_2 is a direct sum of JBW algebras of type $I_{2,k}$.*

Proof. Let M be a JBW algebra of type I_2, and let p be an Abelian projection in M with $c(p)=1$. Let $s=2p-1$. If t is a symmetry in M which interchanges p and $p^\perp$, cf. 5.3.2, then s and t anticommute. Let $(s_\alpha)_{\alpha\in J}$ be a maximal spin system containing s and t.

Let $(e_\beta)_{\beta\in I}$ be a maximal orthogonal family of central projections in M such that the family $(e_\beta s_\alpha)_{\alpha\in J}$ is not a maximal spin system in $e_\beta M$. For each $\beta\in I$ choose a symmetry t_β in $e_\beta M$ which anticommutes with all $e_\beta s_\alpha$, $\alpha\in J$. Let $e=\sum_{\beta\in I} e_\beta$. If $e=1$ then $t=\sum_{\beta\in I} t_\beta$ is a symmetry in M which anticommutes with all s_α, contradicting the maximality of $(s_\alpha)_{\alpha\in J}$. Therefore $e\neq 1$, and by maximality of $(e_\beta)_{\beta\in I}$, $(e^\perp s_\alpha)_{\alpha\in J}$ is a locally maximal spin system in $e^\perp M$. Thus M admits a nonzero summand of type $I_{2,n}$, where $n=\text{card } J$ or $n=\infty$ if card $J\geq\aleph_0$. The proof is completed by the usual Zorn's lemma argument.

6.4. JBW algebras of type I_n, $3\leq n<\infty$

6.4.1. Theorem. *Let M be a JBW algebra of type I_n, $3\leq n<\infty$. Then M is a direct sum $M=M_1\oplus M_2\oplus M_3\oplus M_4$, where $M_4=0$ if $n\neq 3$, such that* *all factor representations of* M_i *are onto* JBW *factors isomorphic to* $H_n(R_i)$*, where* $R_1 = \mathbb{R}$, $R_2 = \mathbb{C}$, $R_3 = \mathbb{H}$ *and* $R_4 = \mathbb{O}$.

Proof. Let π be a factor representation of M. By 5.3.12 $\pi(M)$ is a JBW factor of type I_n, hence is isomorphic to one of the $H_n(R_i)$ by 5.3.8, and $i \neq 4$ if $n \neq 3$. Let $p_1, \ldots, p_n$ be Abelian projections in M with central supports 1 and sum 1. Let $p = p_1 + p_2$. Then $\{pMp\}$ is a JBW algebra of type I_2, and so by 5.2.17 $\{\pi(p)\pi(M)\pi(p)\}$ is a JBW factor of type I_2 isomorphic to $H_2(R_i)$. In particular, $\pi(\{pMp\}) \simeq V_k$ where $k+1 = \dim V_k = 2 + \dim R_i$, hence $k \in \{1+2^l : l = 0, 1, 2, 3\}$. By 6.3.11 and 6.3.14 $\{pMp\}$ is a direct sum $N_1 \oplus \ldots \oplus N_4$, where N_l is the summand of type $\mathrm{I}_{2,k}$ with $k = 1 + 2^{l-1}$. Hence each factor representation of N_l is onto $H_2(R_l)$. Since $c(p) = 1$ the map $Z \to Zp$ is an isomorphism (4.3.4), where Z is the centre of M. Therefore, if e_l is the central projection in $\{pMp\}$ such that $N_l = e_l\{pMp\}$, then there is by 5.2.17 a unique projection $f_l \in Z$ such that $e_l = f_l p$, and $\sum_1^4 f_l = 1$. Let $M_l = f_l M$. Then $M = M_1 \oplus \ldots \oplus M_4$, and each factor representation of M_l is such that $\pi(\{e_l M_l e_l\}) = \pi(N_l) \simeq H_2(R_l)$. Since any factor representation of M_l is onto some $H_n(R_i)$, counting dimensions shows that $i = l$. Thus $\pi(M_l) \simeq H_n(R_l)$.

6.4.2. Corollary. *Let* M *be a* JBW *algebra with no nonzero representations as a* JC *algebra. Then* M *is of type* I_3*, and all its factor representations are onto* $H_3(\mathbb{O})$.

Proof. By 5.3.9 and 6.3.10 M is of type I_3, so by 6.4.1 $M = M_1 \oplus \ldots \oplus M_4$ with all factor representations of M_i onto $H_3(R_i)$, $i = 1, \ldots, 4$ and R_i as in the theorem. Since $H_3(R_i)$ is a JC algebra for $i \leq 3$ (3.1.2), $M = M_4$, proving the result.

6.4.3. Remark. By similar techniques as in 6.3.13 we can write the summands $M_1, \ldots, M_4$ in 6.4.1 as continuous functions with values in $H_n(R_i)$. Then we find $M_i \simeq C(X_i, H_n(R_i))$, $i = 1, \ldots, 4$, where X_i is a compact Hausdorff space, and R_i as in 6.4.1.

6.5. Comments

Infinite-dimensional spin factors were first studied as JC and JW algebras by Topping [110] and then Størmer [103, 105]. Topping [111] then introduced the abstract definition from which he deduced 6.1.5. Theorem 6.1.8 is due to Størmer [103].

The results in Section 6.2 are of 'folklore' character, except 6.2.5 on spin factors of low dimensions, which is due to Hanche-Olsen [57], Robertson [84] and partly by Størmer [103].

The decomposition result 6.3.14 for JBW algebras of type I_2 is due to Stacey [97]. His proof is different from ours in that, instead of using global techniques as we did, he worked with L^∞ functions in a way related to what we indicated in 6.3.13. Theorem 6.4.1 is also due to Stacey [99], again formulated in terms of L^∞ functions.

7

Structure theory

7.1. The universal algebras

7.1.1. In the present chapter we shall prove the main structure theorems for JB and JBW algebras and then show some applications. It turns out that they are as close to finite-dimensional formally real Jordan algebras as one could hope for. In the applications special emphasis will be given to type I algebras, because they include the essential differences from the self-adjoint part of von Neumann algebras. We shall, however, conclude with a section on JBW algebras with modular projection lattice, just to indicate their relationship to finite von Neumann algebras.

The quickest way to prove our results will be through the use of universal associative algebras corresponding to real Jordan algebras. But first recall our convention that if A and B are abstractly defined algebras then a homomorphism $\phi: A \to B$ is a linear map which preserves the relevant multiplicative structure.

7.1.2. Let A be a real Jordan algebra. A *universal specialization* of A is a pair $(\mathcal{U}, u)$, where $\mathcal{U}$ is an associative real algebra and u a homomorphism of A into $\mathcal{U}^J$ such that:

(i) $u(A)$ generates $\mathcal{U}$ as an algebra.

(ii) If B is a real associative algebra and $\phi: A \to B^J$ is a homomorphism then there exists a homomorphism $\hat{\phi}: \mathcal{U} \to B$ such that $\phi = \hat{\phi}u$.

We shall often write $\mathcal{U}$ without mentioning u, and will call $\mathcal{U}$ the *universal associative algebra* for A.

7.1.3. Theorem. *Let A be a real Jordan algebra. Then there exists up to isomorphism a unique universal specialization of A.*

Proof. We first show uniqueness. Suppose $(\mathcal{U}, u)$ and $(\mathcal{V}, v)$ are universal specializations of A. Then there exist by (ii) homomorphisms $\hat{v}: \mathcal{U} \to \mathcal{V}$ and $\hat{u}: \mathcal{V} \to \mathcal{U}$ such that $v = \hat{v}u$ and $u = \hat{u}v$. Thus $u = \hat{u}\hat{v}u$ and $v = \hat{v}\hat{u}v$. By

(i) $\hat{u}\hat{v}$ is the identity of $\mathcal{U}$ and $\hat{v}\hat{u}$ is the identity on $\mathcal{V}$. It follows that $\hat{u}=\hat{v}^{-1}$, proving uniqueness.

To show existence denote by $\overset{n}{\otimes} A$ the tensor product of A with itself n times considered as a real vector space. Let $T(A)$ be the direct sum

$$T(A)=\bigoplus_{n=1}^{\infty} \overset{n}{\otimes} A.$$

Then $T(A)$ is a real associative algebra with $\otimes$ as multiplication. Let I denote the ideal generated by expressions $a\otimes b+b\otimes a-2a\circ b$, $a, b\in A$, let $\mathcal{U}=T(A)/I$, and let $u: A\to\mathcal{U}$ by $u(a)=a+I$. Then u is a homomorphism into $\mathcal{U}$, and $u(A)$ generates $\mathcal{U}$ as an algebra. We show the universal property (ii).

Suppose B is an associative real algebra and $\phi: A\to B^J$ a homomorphism. Define $\phi_1: T(A)\to B$ by linearity and the formula

$$\phi_1(a_1\otimes\ldots\otimes a_n)=\phi(a_1)\ldots\phi(a_n).$$

Then ϕ_1 is a homomorphism, and since

$$\phi_1(a\otimes b+b\otimes a-2a\circ b)=\phi(a)\phi(b)+\phi(b)\phi(a)-2\phi(a\circ b)=0,$$

ϕ_1 annihilates I. Thus ϕ_1 induces a homomorphism $\hat{\phi}: \mathcal{U}\to B$ such that $\phi=\hat{\phi}\circ u$.

7.1.4. If $\mathcal{A}$ is an associative algebra its *opposite algebra* $\mathcal{A}^0$ is defined to be equal to $\mathcal{A}$ as a vector space, but its multiplication is given by $(a, b)\to ba$. We write $a\to a^0$ for the identity map $\mathcal{A}\to\mathcal{A}^0$, so that $a^0b^0=(ba)^0$ holds.

7.1.5. Lemma. *Let A be a real Jordan algebra and let $(\mathcal{U}, u)$ be its universal specialization. Then there exists an involution $*$ on $\mathcal{U}$ such that $u(a)^*=u(a)$ for $a\in A$. Furthermore, if A is unital then so is $\mathcal{U}$.*

Proof. The homomorphism $u: A\to\mathcal{U}^J$ defines a homomorphism $u^0: A\to(\mathcal{U}^0)^J$ of A into the opposite algebra of $\mathcal{U}$ by $u^0(a)=u(a)^0$ for $a\in A$. By the universal property of $\mathcal{U}$ there is a homomorphism $\hat{u}^0: \mathcal{U}\to\mathcal{U}^0$ such that $\hat{u}^0u=u^0$. Define x^* by $(x^*)^0=\hat{u}^0(x)$ for $x\in\mathcal{U}$. Then $*$ is the desired involution. Indeed, if $x, y\in\mathcal{U}$ then $((xy)^*)^0=\hat{u}^0(xy)=\hat{u}^0(x)\hat{u}^0(y)=(x^*)^0(y^*)^0=(y^*x^*)^0$. For $a\in A$ we get $(u(a)^*)^0=\hat{u}^0u(a)=u^0(a)=u(a)^0$, so $u(a)^*=u(a)$. Then $x\to x^{**}$ is an automorphism of $\mathcal{U}$ which is the identity on $u(A)$. Since A generates $\mathcal{U}$, $x^{**}=x$ follows. Notice that we could have defined the involution from the construction of $\mathcal{U}$ from $T(A)$, and defining involution in $T(A)$ by $(a_1\otimes\ldots\otimes a_n)^*=a_n\otimes\ldots\otimes a_1$.

If A is unital with identity 1 then $e=u(1)$ is an idempotent in $\mathcal{U}$ such that $u(a)=u(\{1a1\})=eu(a)e$ for all $a\in A$. Since $u(A)$ generates $\mathcal{U}$ as an algebra e is the identity in $\mathcal{U}$.

7.1.6. Let $\mathcal{U}^{\mathbb{C}}=\mathcal{U}\oplus i\mathcal{U}$ be the complexification of $\mathcal{U}$. Define $(x+iy)^*=x^*-iy^*$. Then $*$ is an involution on $\mathcal{U}^{\mathbb{C}}$ which extends that of $\mathcal{U}$. The map

$$\Phi(x+iy)=x^*+iy^*$$

is a complex linear $*$ antiautomorphism of $\mathcal{U}^{\mathbb{C}}$ of order 2. If $a\in A$ then $u(a)=u(a)^*=\Phi(u(a))$.

7.1.7. Proposition. *Let A be a real Jordan algebra and let B be a complex associative algebra with involution $*$. Suppose $\phi: A\to B_{sa}$ (the self-adjoint elements in B) is a homomorphism. Then there exists a unique $*$ homomorphism $\hat{\phi}: \mathcal{U}^{\mathbb{C}}\to B$ such that $\phi=\hat{\phi}u$.*

Proof. This is immediate from the preceding discussion.

7.1.8. Theorem. *Let A be a JB algebra. Then there exist up to isomorphism a unique C^* algebra $C_u^*(A)$ and a homomorphism $\psi_A: A\to C_u^*(A)_{sa}$ such that:*

(i) *$\psi_A(A)$ generates $C_u^*(A)$ as a C^* algebra.*
(ii) *If $\mathcal{A}$ is a C^* algebra and $\phi: A\to \mathcal{A}_{sa}$ is a homomorphism then there exists a $*$ homomorphism $\hat{\phi}: C_u^*(A)\to\mathcal{A}$ such that $\phi=\hat{\phi}\psi_A$.*
(iii) *There is a $*$ antiautomorphism Φ of $C_u^*(A)$ of order 2 such that $\Phi(\psi_A(a))=\psi_A(a)$ for all $a\in A$.*

Proof. Let $(\mathcal{U}^{\mathbb{C}}, u)$ denote the complexified universal specialization of A. If $x\in\mathcal{U}^{\mathbb{C}}$ let

$$\|x\|=\sup\{\|\pi(x)\|: \pi \text{ a } * \text{ representation of } \mathcal{U}^{\mathbb{C}}\}.$$

Then $\|x\|<\infty$. Indeed, x is an algebraic combination of elements in $u(A)$, hence it suffices to show $\|u(a)\|<\infty$ for all $a\in A$. But if π is a $*$ representation of $\mathcal{U}^{\mathbb{C}}$ on a Hilbert space H, then πu is a homomorphism of A into $B(H)_{sa}$, hence is norm decreasing by 3.4.3. Thus $\|\pi(u(a))\|\leq\|a\|$ for all π, and so $\|u(a)\|\leq\|a\|<\infty$. Clearly $\|\ \ \|$ has all the other properties of a seminorm on $\mathcal{U}^{\mathbb{C}}$, and moreover it is a C^* seminorm on $\mathcal{U}^{\mathbb{C}}$, viz. $\|x^*x\|=\|x\|^2$ (1.3.2). Let $N=\{x\in\mathcal{U}: \|x\|=0\}$. Then N is a two-sided ideal in $\mathcal{U}^{\mathbb{C}}$ such that $x\in N$ implies $x^*\in N$. Let $C_u^*(A)$ be the completion of $\mathcal{U}^{\mathbb{C}}/N$ in the quotient norm. Then $C_u^*(A)$ is a C^* algebra, and if $\psi_A(a)=u(a)+N$ for $a\in A$ then (i) is immediate.

To show (ii) let $\phi: A\to\mathcal{A}_{sa}$ be a homomorphism. Let by 7.1.7 ϕ_1 be the $*$ homomorphism of $\mathcal{U}^{\mathbb{C}}$ into $\mathcal{A}$ such that $\phi_1 u=\phi$. Then in particular

$\pi\phi_1$ is a $*$ representation of $\mathcal{U}^{\mathrm{C}}$ for all $*$ representations π of $\mathcal{A}$ so if $x \in N$ then $\|\pi\phi_1(x)\| = 0$, hence $\|\phi_1(x)\| = 0$. Thus ϕ_1 induces a $*$ homomorphism $\hat{\phi}: C_u^*(A) \to \mathcal{A}$ such that $\hat{\phi}\psi_A = \phi$.

To show (iii) let Φ' be the $*$ antiautomorphism of $\mathcal{U}^{\mathrm{C}}$ defined by 7.1.6. Suppose $x \in N$ and let π be a $*$ representation of $\mathcal{U}^{\mathrm{C}}$ on a Hilbert space H. Let α be any $*$ antiautomorphism of $B(H)$ (α exists by 7.5.6 below). Then $\alpha\pi\Phi'$ is a $*$ representation of $\mathcal{U}^{\mathrm{C}}$ on H, hence $\|\pi(\Phi'(x))\| = \|\alpha\pi\Phi'(x)\| = 0$, and $\Phi'(x) \in N$. Thus Φ' induces a $*$ antiautomorphism Φ of order 2 of $\mathcal{U}^{\mathrm{C}}/N$, hence of $C_u^*(A)$. Clearly $\Phi(\psi_A(a)) = \psi_A(a)$ for all $a \in A$.

The proof of uniqueness is a trivial modification of the uniqueness proof in 7.1.3 and is omitted.

7.1.9. Theorem. *Let M be a* JBW *algebra. Then there exist up to isomorphism a unique von Neumann algebra* $W^*(M)$ *and a normal homomorphism $\psi: M \to W^*(M)$ such that:*

(i) *$\psi(M)$ generates $W^*(M)$ as a von Neumann algebra.*

(ii) *If $\mathcal{N}$ is a von Neumann algebra and $\phi: M \to \mathcal{N}_{\mathrm{sa}}$ is a normal homomorphism then there is a normal $*$ homomorphism $\hat{\phi}: W^*(M) \to \mathcal{N}$ such that $\hat{\phi}\psi = \phi$.*

(iii) *There is a $*$ antiautomorphism Φ of order* 2 *of $W^*(M)$ such that $\Phi(\psi(a)) = \psi(a)$ for all $a \in M$.*

Proof. Let $\iota: C_u^*(M) \to C_u^*(M)^{**}$ denote the inclusion map, and let $\psi_M: M \to C_u^*(M)$ be the homomorphism found in 7.1.8. Let (e_α) be a maximal orthogonal family of central projections in $C_u^*(M)^{**}$ such that the map $M \to e_\alpha C_u^*(M)^{**}$ defined by $a \to e_\alpha(\iota\psi_M(a))$ is normal. Let $e = \sum_\alpha e_\alpha$. Then e is the maximal central projection with this property. We let $W^*(M) = eC_u^*(M)^{**}$ and $\psi: A \to W^*(M)$ be the map $\psi(a) = e(\iota\psi_M(a))$. By 2.5.6 ψ is a homomorphism, hence (i) follows.

Let $\mathcal{N}$ be a von Neumann algebra and $\phi: M \to \mathcal{N}_{\mathrm{sa}}$ be a normal homomorphism. Let ϕ_1 denote the extension of ϕ to $C_u^*(M)$ such that $\phi_1\psi_M = \phi$. Let $\bar{\phi}: C_u^*(M)_{\mathrm{sa}}^{**} \to \mathcal{N}_{\mathrm{sa}}$ be the normal extension of ϕ_1 to $C_u^*(M)_{\mathrm{sa}}^{**}$ found in 4.5.7, and extend $\bar{\phi}$ by linearity to all of $C_u^*(M)^{**}$. We thus have the following commutative diagram.

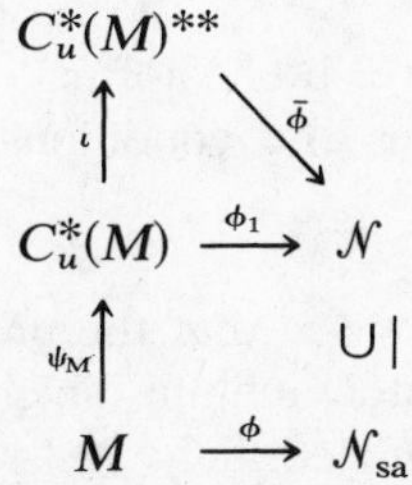

$$
\begin{array}{ccc}
C_u^*(M)^{**} & & \\
\uparrow{\scriptstyle \iota} & \searrow{\scriptstyle \bar{\phi}} & \\
C_u^*(M) & \xrightarrow{\phi_1} & \mathcal{N} \\
\uparrow{\scriptstyle \psi_M} & & \cup\!| \\
M & \xrightarrow{\phi} & \mathcal{N}_{\mathrm{sa}}
\end{array}
$$

Let f be the support of $\bar{\phi}$ in $C_u^*(M)^{**}$, i.e. f is by 4.3.7 the central projection such that $\bar{\phi}(1-f)=0$ and $\bar{\phi} \mid fC_u^*(M)^{**}$ is injective. By commutativity of the diagram the map $M \to fC_u^*(M)^{**}$ by $a \to f(\iota\psi_M(a))$ decomposes as

$$a \to \phi(a) \to (\bar{\phi} \mid fC_u^*(M)^{**})^{-1}(\phi(a)),$$

which is a composition of normal maps, hence is normal. Thus $f \leq e$ by maximality of e, so that $\hat{\phi} = \bar{\phi} \mid W^*(M)$ satisfies (ii).

To show (iii) let by 7.1.8 Φ' denote the $*$ antiautomorphism of order 2 of $C_u^*(M)$ such that $\Phi'(\psi_M(a)) = \psi_M(a)$ for all $a \in M$. Then Φ' has an extension Φ'' to $C_u^*(M)^{**}$, which by 1.1.21 and 4.5.7 is a normal $*$ antiautomorphism of order 2. If e is as above then the map $M \to \Phi''(e)C_u^*(M)^{**} = \Phi''(eC_u^*(M)^{**})$ is a normal map, hence by maximality of e, $\Phi''(e) \leq e$. Since Φ'' is of order 2 we have equality, and $\Phi = \Phi'' \mid eC_u^*(M)^{**}$ is the desired $*$ antiautomorphism.

As in 7.1.8 uniqueness is trivial.

7.1.10. The $*$ antiautomorphism Φ of 7.1.8 (resp. 7.1.9) is called the *canonical antiautomorphism* of $C_u^*(A)$ (resp. $W^*(M)$).

7.1.11. Proposition. *For any* JB *algebra A we have a natural isomorphism $W^*(A^{**}) \cong C_u^*(A)^{**}$.*

Proof. We must show that $C_u^*(A)^{**}$ with the homomorphism $\psi_A^{**}: A^{**} \to C_u^*(A)^{**}$ satisfies the defining properties of $W^*(A^{**})$. Since $\psi_A(A)$ generates $C_u^*(A)$ as a C^* algebra, it generates $C_u^*(A)^{**}$ as a von Neumann algebra. But $\psi_A^{**}(A^{**})$ contains $\psi_A(A)$, hence it too generates $C_u^*(A)^{**}$.

Let $\phi: A^{**} \to \mathcal{N}_{sa}$ be a normal homomorphism, where $\mathcal{N}$ is a von Neumann algebra. Let ϕ_0 be the restriction of ϕ to A, and consider the $*$ homomorphism $\hat{\phi}_0: C_u^*(A) \to \mathcal{N}$ such that $\phi_0 = \hat{\phi}_0\psi_A$ (7.1.8). By 4.5.7, $\hat{\phi}_0$ admits a normal extension $\hat{\phi}: C_u^*(A)^{**} \to \mathcal{N}$. Since the product in $C_u^*(A)^{**}$ is separately weakly continuous, $\hat{\phi}$ is actually a $*$ homomorphism. On A we have

$$\hat{\phi}\psi_A^{**} \mid_A = \hat{\phi}_0\psi_A = \phi_0 = \phi \mid_A,$$

and since $\hat{\phi}\psi_A^{**}$ and ϕ are both normal, $\hat{\phi}\psi_A^{**} = \phi$ follows. Thus $\psi_A^{**}: A^{**} \to C_u^*(A)_{sa}^{**}$ satisfies the conditions of 7.1.9, so $C_u^*(A)^{**} = W^*(A^{**})$.

7.1.12. Remark. We can now see that the universal C^* algebras of the finite-dimensional and separately infinite-dimensional spin factors are the algebras exhibited in 6.2.2. Indeed, if $2 \leq k < \infty$ then $C_u^*(V_k)$ is generated by k anticommuting symmetries $s_1, \ldots, s_k$. It is then easy to see that the 2^k elements of the form $s_{i_1} \ldots s_{i_j}$ where $1 \leqslant i_1 < i_2 < \ldots < i_j \leqslant k$ and $0 \leqslant j \leqslant k$ generate $C_u^*(V_k)$ linearly. Indeed, the product of any two such elements is a third such modulo sign. For example, $(s_1s_3s_5)(s_1s_4) = -s_3s_4s_5$. Thus $\dim C_u^*(V_k) \leqslant 2^k$. However, the algebra $C^*(V_k)$ defined in 6.2.2 is generated by V_k, and by 7.1.8 there is a homomorphism of $C_u^*(V_k)$ onto $C^*(V_k)$. But $\dim C^*(V_k) = 2^k$, so this homomorphism must be an isomorphism. The corresponding statement for $k = \infty$, that $C_u^*(V_\infty)$ is the CAR algebra, follows since it is easy to show that C_u^* is a functor preserving inductive limits.

7.2. Structure of JB and JBW algebras

7.2.1. Let A be a JB algebra. We shall call A *purely exceptional* if there is no nonzero homomorphism of A into a JC algebra. Since every JBW factor other than $H_3(\mathbb{O})$ is a JW algebra (5.3.8, 5.3.9, 6.2.3), it follows that every factor representation of A is onto $H_3(\mathbb{O})$.

7.2.2. Lemma. *A* JB *algebra A is purely exceptional if and only if every factor representation of A is onto* $H_3(\mathbb{O})$.

Proof. It remains to prove the 'if' part. Assume that A is not purely exceptional. Then there is a nonzero homomorphism ϕ of A onto a JC algebra B. By 4.6.4 there is a factor representation ψ of B. By 4.7.4 $\psi(B)$ is a JC algebra, so, since $H_3(\mathbb{O})$ is exceptional (2.8.5), $\psi(B)$ is not (isomorphic to) $H_3(\mathbb{O})$. Hence $\psi\phi$ is a factor representation of A not onto $H_3(\mathbb{O})$.

7.2.3. Theorem. *Any* JB *algebra A contains a unique purely exceptional ideal J such that A/J is a* JC *algebra.*

Proof. Let by 7.1.8 $C_u^*(A)$ be the universal C^* algebra for A, and let ψ_A denote the canonical homomorphism of A into $C_u^*(A)$. Let J be the kernel of ψ_A. Then A/J is, by 3.4.2 and 3.4.3, isometrically isomorphic to a norm-closed Jordan subalgebra of $C_u^*(A)_{sa}$, and hence is a JC algebra. If J is not purely exceptional there is a nonzero homomorphism of J into a JC algebra, and hence into $B(H)_{sa}$. By 4.5.8 this extends to a homomorphism $\pi: A \to B(H)_{sa}$. But then by 7.1.8 there is a * homomorphism $\hat{\pi}$ of $C_u^*(A)$ into $B(H)_{sa}$ such that $\pi = \hat{\pi}\psi_A$. In particular π vanishes on J, which is a contradiction. Thus J is purely exceptional.

To prove uniqueness let K be another purely exceptional ideal. Then ψ_A must vanish on K, so that $K \subseteq J$. If A/K is a JC algebra then there is a homomorphism $\pi: A \to B(H)_{sa}$ with kernel K. Then as above π vanishes on J, so $J \subseteq K$. The proof is complete.

7.2.4. Recall that the Shirshov–Cohn theorem (2.4.14) states that a Jordan algebra, unital or not, with two generators is special. As a consequence of 7.2.3 we are now able to prove the JB version of this theorem.

7.2.5. The Shirshov–Cohn theorem for JB algebras. *A* JB *algebra generated by two elements is a* JC *algebra.*

Proof. Let A be a JB algebra with two generators a and b, and possibly an identity 1. If A is not a JC algebra A has by 7.2.2, 7.2.3 and 4.5.8 a representation π on $H_3(\mathbb{O})$. Then $\pi(A)$ is finite-dimensional, so is generated algebraically by $\pi(a)$ and $\pi(b)$, and possibly the identity $\pi(1)$. But then by the Shirshov–Cohn theorem (2.4.14) $\pi(A)$ is special, contradicting the fact that $H_3(\mathbb{O})$ is exceptional (2.8.5).

7.2.6. Remark. It might be expected that 7.2.3 could be improved in the sense that A is the direct sum of A/J and J. This is not true, as the following example shows.

For $n \in \mathbb{N}$ let $A_n = H_3(\mathbb{O})$, and let $A \subset \bigoplus_{n=1}^{\infty} A_n$ consist of all convergent sequences (a_n) where $a_n = (x_{ij}^n)_{i,j=1,2,3} \in H_3(\mathbb{O})$, and $x_{ij}^n \to 0$ for $i \neq j$. Then it is easy to show that with pointwise operations A is a JB algebra, $J = \{(a_n): a_n \to 0\}$, and A/J is three-dimensional and associative.

If A is a JBW algebra, however, we next prove that A is indeed the direct sum of A/J and J.

7.2.7. Theorem. *Let M be a* JBW *algebra. Then M can be uniquely decomposed as a direct sum $M = M_{ex} \oplus M_{sp}$, where M_{sp} is a* JW *algebra and M_{ex} is a purely exceptional* JBW *algebra.*

Proof. By 5.1.5 and 5.3.5 M has a unique decomposition

$$M = M_{I_1} \oplus M_{I_2} \oplus \ldots \oplus M_{I_\infty} \oplus M_{II} \oplus M_{III},$$

where each summand is of the type indicated. By 5.3.9 and 6.3.10 each summand other than M_{I_3} is a JW algebra. By 6.4.1 $M_{I_3} = M_1 \oplus \ldots \oplus M_4$ where all factor representations of M_i are onto $H_3(R_i)$, where $R_1 = \mathbb{R}$, $R_2 = \mathbb{C}$, $R_3 = \mathbb{H}$, $R_4 = \mathbb{O}$. Let $M_{ex} = M_4$ and M_{sp} be the direct sum of the rest of the summands. Then M_{sp} is a JW algebra, and M_{ex} is purely exceptional by 7.2.2.

7.2.8. Remark. Let by 7.1.9 ψ be the canonical homomorphism of M into $W^*(M)$. If $M = M_{ex} \oplus M_{sp}$ as in 7.2.7 then $M_{ex} = \ker \psi$. Indeed, let

$J = \ker \psi$. Then J is a weakly closed ideal in M, so by 4.3.6 there is a central projection e in M such that $J = eM$. Let p and q be the central projections in M such that $M_{ex} = pM$, $M_{sp} = qM$. If $J \cap M_{sp} \neq \{0\}$ then the homomorphism $a \to qea$ is a normal representation of M as a JW algebra not annihilating the kernel of ψ. This contradicts 7.1.9(ii), so that $eq = 0$. Thus $e \leq p$. If $e \neq p$ then $a \to (1-e)pa$ is a representation of M annihilating J, hence it induces a representation of $(1-e)M_{ex}$ into $\psi(M)$. Since M_{ex} has no representation as a JC algebra while $\psi(M)$ is a JC algebra, this is impossible. Thus $e = p$, proving the assertion.

7.3. Antiautomorphisms

7.3.1. The theory of $*$ antiautomorphisms of von Neumann algebras, and especially those of order 2, is intimately related to the theory of reversible JW algebras. Indeed, if α is a $*$ antiautomorphism of order 2 of the von Neumann algebra $\mathcal{M}$, the reader should have no difficulty in showing that the self-adjoint part of the set $\mathcal{M}^{\alpha}$ of fixed points under α is a reversible JW algebra. We shall in the present section develop some of this theory with particular emphasis on the canonical antiautomorphism Φ of $W^*(M)$ (7.1.10). Note that if M is a JW algebra then $M_{ex} = 0$, cf. 7.2.7, so if ψ is the canonical homomorphism of M into $W^*(M)$ exhibited in 7.1.9, then ψ is injective by 7.2.8. We may therefore consider M as a JW subalgebra of $W^*(M)_{sa}$ in this case. By 7.1.9 $\Phi(a) = a$ for all $a \in M$, hence $M \subset W^*(M)_{sa}^{\Phi}$. We shall shortly show that the converse inclusion holds in most cases.

7.3.2. Lemma. *Let $\mathcal{M}$ be a von Neumann algebra and α a $*$ antiautomorphism of $\mathcal{M}$ of order 2. Let $R = \{z \in \mathcal{M}: \alpha(z) = z^*\}$. Then R is an ultraweakly closed real $*$ algebra; $\mathcal{M}$ is the direct sum $\mathcal{M} = R \oplus iR$, and $\alpha(x + iy) = x^* + iy^*$, $x, y \in R$.*

Proof. Since the $*$ operation is ultraweakly continuous, multiplication is separately ultraweakly continuous (cf. 1.4.1) and α is normal (4.5.6), it is straightforward to show that R is an ultraweakly closed real $*$ subalgebra of $\mathcal{M}$. If $z \in \mathcal{M}$ then $z = \frac{1}{2}(z + \alpha(z^*)) + i(1/2i)(z - \alpha(z^*)) \in R + iR$. If $x = iy \in R \cap iR$ then $x^* = \alpha(x) = i\alpha(y) = iy^*$, hence $x = -iy$, so $x = y = 0$. Therefore the sum is direct. The last statement follows by definition of R.

7.3.3. Proposition. *Let M be a JW algebra with no type I_2 part. Then $M = W^*(M)_{sa}^{\Phi}$, where Φ is the canonical antiautomorphism of $W^*(M)$.*

Proof. By 7.3.2 $W^*(M) = R + iR$ is a direct sum, where R is an ultraweakly closed real $*$ algebra, and $\Phi(x + iy) = x^* + iy^*$, $x, y \in R$. We show $W^*(M)^\Phi_{sa} \subset M$, which by 7.3.1 will prove the proposition. Let $z \in W^*(M)^\Phi_{sa}$. By 7.3.2 $z \in R_{sa}$. Since M generates $W^*(M)$ as a von Neumann algebra, cf. 7.3.1, z is the ultraweak limit of self-adjoint operators $x + iy$ with x, y in the real algebra generated by M. In particular $x, y \in R$, since $M \subset W^*(M)^\Phi_{sa}$ by 7.3.1. Furthermore since $x + iy$ is self-adjoint, $x = x^*$, $y = -y^*$. Since $z = \Phi(z)$, we may assume $\Phi(x + iy) = x + iy$, i.e. $x^* + iy^* = x + iy$, so $y = 0$. Now x is a finite sum of finite products, $x = \sum a_1 \dots a_k$, with $a_1, \dots, a_k \in M$. Since $x = \frac{1}{2}(x + x^*) = \frac{1}{2}\sum (a_1 \dots a_k + a_k \dots a_1)$, $x \in M$, since by 5.3.10 M is reversible. Thus z, being an ultraweak limit of elements in M belongs to M, and so $W^*(M)^\Phi_{sa} \subset M$, proving the proposition.

7.3.4. Lemma. *Let $\mathscr{Z}$ be an Abelian von Neumann algebra and Φ a $*$ automorphism of order 2 of $\mathscr{Z}$. Then there exist projections e and f in $\mathscr{Z}$ such that $e + f + \Phi(f) = 1$, and $\Phi(g) = g$ for all projections $g \leq e$.*

Proof. Let by Zorn's lemma (f_α) be a maximal family of projections in $\mathscr{Z}$ such that $\sum_\alpha (f_\alpha + \Phi(f_\alpha)) \leq 1$. Then all f_α and $\Phi(f_\beta)$ are pairwise orthogonal, so if $f = \sum_\alpha f_\alpha$ and $e = 1 - f - \Phi(f)$, then e and f are projections in $\mathscr{Z}$. Suppose g is a nonzero projection majorized by e. Let $f' = g(1 - \Phi(g))$. Then $f' \leq g$, and $\Phi(f') = \Phi(g)(1 - g) \leq \Phi(e)(1 - g) = e - g$, so that $f' \perp \Phi(f')$ and $f' + \Phi(f') \leq e$, contradicting the maximality of (f_α) unless $f' = 0$. Therefore $0 = f' = g - g\Phi(g)$, which shows $g \leq \Phi(g)$. Since Φ is of order 2, $\Phi(g) = g$.

7.3.5. Theorem. *Let M be a JW algebra with no direct summand of type I_2. Then there exists a central projection e in M such that the centre of $eW^*(M)e$ $(= W^*(eM))$ is pointwise Φ-invariant for the canonical anti-automorphism Φ, and such that $(1 - e)M$ is isomorphic to the self-adjoint part of a von Neumann algebra.*

Proof. Let $\mathscr{Z}$ denote the centre of $W^*(M)$. From 7.3.4 there are projections e and f in $\mathscr{Z}$ such that $e + f + \Phi(f) = 1$, and $\Phi(g) = g$ for all projections $g \in \mathscr{Z}$, $g \leq e$. In particular $\mathscr{Z}e$ is pointwise Φ-invariant.

Let $\mathscr{N} = fW^*(M)$. Then $\mathscr{N}$ is a von Neumann algebra, and Φ is an anti-isomorphism of $\mathscr{N}$ onto $\Phi(f)W^*(M)$. In particular $\Phi(f)W^*(M)$ is isomorphic to the opposite algebra $\mathscr{N}^0$ of $\mathscr{N}$. Therefore we have

$$W^*(e^\perp M) = e^\perp W^*(M) = fW^*(M) + \Phi(f)W^*(M) = \mathscr{N} \oplus \mathscr{N}^0.$$

With this identification of $\Phi(\mathscr{N})$ and $\mathscr{N}^0$, Φ corresponds to the anti-automorphism of $\mathscr{N} \oplus \mathscr{N}^0$ defined by $a \oplus b^0 \to b \oplus a^0$, $a, b \in \mathscr{N}$, where $a \to a^0$ is the identity map of $\mathscr{N}$ onto $\mathscr{N}^0$. By 7.3.3 we have

$$e^\perp M = (\mathscr{N} \oplus \mathscr{N}^0)^\Phi_{sa} = \{a \oplus a^0 \colon a \in \mathscr{N}_{sa}\},$$

which is isomorphic to $\mathscr{N}_{sa}$ via the map $a \oplus a^0 \to a$.

7.3.6. Remark. We showed above that there exist central projections e and f in $W^*(M)$ such that $e+f+\Phi(f)=1$, and $\Phi(g)=g$ for all central projections g majorized by e. We used that M has no type I_2 part in order to apply 7.3.3, or rather to conclude that M is reversible. Thus the theorem is true if M has a reversible I_2 part. It follows that if M is reversible in $W^*(M)$ then $W^*(e^{\perp}M)\cong\mathcal{N}\oplus\mathcal{N}^0$.

7.3.7. Let $\mathcal{M}$ be a von Neumann algebra with centre $\mathcal{Z}$. A linear map $E\colon\mathcal{M}\to\mathcal{Z}$ is called a *centre-valued trace* if $x\geqslant 0$ implies $E(x)\geqslant 0$, $E(1)=1$, $E^2=E$ and $E(xy)=E(yx)$ for all $x, y\in\mathcal{M}$. If moreover $x\geqslant 0$ and $E(x)=0$ imply $x=0$, then E is said to be faithful.

$\mathcal{M}$ is said to be of *type* I_k, II or III if $\mathcal{M}_{sa}$ is of the same type as a JBW algebra. Thus by 5.1.5 and 5.3.5 $\mathcal{M}$ is a direct sum of von Neumann algebras of different types.

7.3.8. Lemma. *Let $\mathcal{M}$ be a von Neumann algebra with centre $\mathcal{Z}$. Suppose Φ is a $*$ antiautomorphism of order 2 of $\mathcal{M}$ such that $\mathcal{M}^{\Phi}\subset\mathcal{Z}$. Then $\mathcal{M}$ decomposes as $\mathcal{M}=\mathcal{M}_1\oplus\mathcal{M}_2$ with $\mathcal{M}_i$ of type I_i.*

Proof. Let $E=\frac{1}{2}(\iota+\Phi)$, where ι denotes the identity map. Then E is a faithful idempotent positive map of $\mathcal{M}$ onto $\mathcal{M}^{\Phi}\subset\mathcal{Z}$. We show E is a centre-valued trace. For this let $x, y\in\mathcal{M}$. Then $E(xy)=\frac{1}{2}[xy+\Phi(y)\Phi(x)]$, so that

$$\begin{aligned}2[E(xy)-E(yx)]&=xy-yx+\Phi(y)\Phi(x)-\Phi(x)\Phi(y)\\&=[xy+x\Phi(y)]-[x\Phi(y)+\Phi(x)\Phi(y)]\\&\quad-[yx+\Phi(y)x]+[\Phi(y)x+\Phi(y)\Phi(x)]\\&=2[xE(y)-E(x)\Phi(y)-E(y)x+\Phi(y)E(x)]\\&=0,\end{aligned}$$

since $E(x)$ and $E(y)$ belong to $\mathcal{Z}$.

Since Φ is an antiautomorphism, Φ leaves all parts of different types of $\mathcal{M}$ invariant. We shall assume $\mathcal{M}$ has no parts of types I_1 and I_2 and shall obtain a contradiction.

Since $\mathcal{M}$ has no parts of types I_1 and I_2 there are (by definition if $\mathcal{M}$ is of type I and by 5.2.15 otherwise) projections p and q in $\mathcal{M}$ with $0\neq p<q$ such that their central supports satisfy $c(p)=c(q-p)=c(1-q)=1$. Clearly $q\vee\Phi(q)$ is Φ-invariant, so belongs to $\mathcal{Z}$. Since $c(q)=1$ and $q\vee\Phi(q)\geqslant q$, $q\vee\Phi(q)=1$. Similarly $p\vee\Phi(p)=1$, so $p\vee\Phi(q)\geqslant p\vee\Phi(p)=1$. Similarly $q^{\perp}\vee\Phi(q)^{\perp}\in\mathcal{Z}$, and is greater than $1-q=q^{\perp}$, which has central support 1. Thus we have $(q\wedge\Phi(q))^{\perp}=q^{\perp}\vee\Phi(q)^{\perp}=1$, and so $q\wedge\Phi(q)=0$. Since $p\leqslant q$ it follows that $p\wedge\Phi(q)=0$. Thus $\Phi(q)$ is a common complement for p and q, so that p and q are perspective (see 5.1.2). By 5.2.3 $p\underset{\sim}{\sim}q$ in the

JBW algebra $\mathcal{M}_{sa}$. Since E is a centre-valued trace $E(sxs)=E((sx)s)=E(ssx)=E(x)$ for all symmetries s in $\mathcal{M}_{sa}$ and $x \in \mathcal{M}$. Thus $E(p)=E(q)$, or $E(q-p)=0$. Since E is faithful and $p \leq q$, $p=q$ contrary to assumption. Thus $\mathcal{M}=\mathcal{M}_1 \oplus \mathcal{M}_2$ as asserted.

7.4. Connections with von Neumann algebras

7.4.1. The interplay between JW algebras and the von Neumann algebras they generate is very important both in applications of JW algebras and in their intrinsic theory. We shall in the present section see some of this interplay, and then use it to obtain some applications to Jordan homomorphisms of C^* algebras.

7.4.2. Theorem. *Let M be a JW algebra with no direct summand of type* I_2. *Then we have:*
(i) *If M is of type* I *then so is* $W^*(M)$.
(ii) *If M has no direct summand of type* I *neither does* $W^*(M)$.

Proof. By 7.3.3 $M=W^*(M)^{\Phi}_{sa}$, where Φ is the canonical antiautomorphism of $W^*(M)$. Let Z denote the centre of M and $\mathscr{Z}$ that of $W^*(M)$. If $\mathcal{N}$ is a von Neumann algebra or a JW algebra we use the notation $c_{\mathcal{N}}(p)$ to denote the central support in $\mathcal{N}$ of a projection p in $\mathcal{N}$.

Suppose M is of type I, and let p be an Abelian projection in M with $c_M(p)=1$. Let $\mathcal{N}=pW^*(M)p$. Then $\mathcal{N}$ is Φ-invariant since $\Phi(p)=p$. Suppose $a \in \mathcal{N}^{\Phi}_{sa}$. Then $a \in W^*(M)^{\Phi}_{sa}=M$, so $a \in \{pMp\}$. By 5.2.17, $a \in Zp$. Since M generates $W^*(M)$ (7.1.9) $Z \subset \mathscr{Z}$ (4.3.8), so that a belongs to the centre of $\mathcal{N}$. Thus $\mathcal{N}$ is of type I by 7.3.8. Let q be an Abelian projection in $\mathcal{N}$ with $c_{\mathcal{N}}(q)=p$. Then q is Abelian in $W^*(M)$, and in order to finish the proof of (i) it remains to show $c_{W^*(M)}(q)=1$. For this let $e \in \mathscr{Z}$ be a projection such that $e \geq q$. Then $ep \geq q$. Now by 5.2.17 the centre of $\mathcal{N}$ is $\mathscr{Z}p$, so ep belongs to the centre of $\mathcal{N}$. Therefore $ep \geq p$, and so $ep=p$. It follows that $e \geq p$ and therefore $c_{W^*(M)}(q)=c_{W^*(M)}(p)$. Since $p \in M$, $\Phi(p)=p$, hence $\Phi(c_{W^*(M)}(p))=c_{W^*(M)}(p)$. Therefore $c_{W^*(M)}(p) \in M \cap \mathscr{Z}=Z$, and $c_{W^*(M)}(p)=c_M(p)=1$. Thus $c_{W^*(M)}(q)=1$ completing the proof of (i).

In order to show (ii) suppose $W^*(M)$ contains a nonzero Abelian projection p. Let $q=p \vee \Phi(p)$. Since by 7.3.3 $M=W^*(M)^{\Phi}_{sa}$ and q is Φ-invariant, $q \in M$. By 5.2.3 we have

$$q-p=p \vee \Phi(p)-p \sim \Phi(p)-p \wedge \Phi(p)$$

in $W^*(M)$. Since $\Phi(p)$ is an Abelian projection in $W^*(M)$ so is each subprojection (5.3.2), and hence $q-p$ is Abelian. Therefore q is the orthogonal sum of two Abelian projections, and so $qW^*(M)q$ is decomposed into a direct sum of a type I_1 and a type I_2 part. In particular qMq can at most contain two orthogonal equivalent projections. By 5.2.15 qMq is of type I, so M has a type I part, proving (ii).

7.4.3. Theorem. *Let M be a JW algebra with no direct summand of type I_2. Suppose M acts on a Hilbert space H, and let $\mathcal{M}$ denote the von Neumann algebra generated by M. Then M is of type I if and only if $\mathcal{M}$ is of type I.*

Proof. By 7.1.9 there is a normal $*$ homomorphism of $W^*(M)$ onto $\mathcal{M}$. Therefore, if M is of type I then by 7.4.2 $\mathcal{M}$ is of type I.

Conversely, assume $\mathcal{M}$ is of type I. Let by 5.1.5 e be the central projection in M such that $(1-e)M$ is of type I and eM has no type I portion. Suppose $e \neq 0$. Since eM generates $e\mathcal{M}$ as a von Neumann algebra there is by 7.1.9 applied to eM a normal $*$ homomorphism ϕ of $W^*(eM)$ onto $e\mathcal{M}$. Since $\mathcal{M}$ is of type I so is $e\mathcal{M}$, and therefore $W^*(eM)$ has a type I portion. By 7.4.2 so has eM contrary to assumption. Thus $e = 0$, proving the theorem.

7.4.4. Remark. The conclusion of the theorem is definitely false if M is of type I_2. Indeed, by 6.2.2 if M is the infinite-dimensional spin factor of countable dimension then the C^* algebra generated by M is the CAR algebra. This C^* algebra is known to have factor representations of types I, II and III. Thus a JW factor of type I_2 can generate von Neumann algebras of all types.

7.4.5. Lemma. *Let $\mathcal{M}$ be a von Neumann algebra of type I_n, $n<\infty$, with centre $\mathcal{Z}$. Then $M \cong M_n(\mathcal{Z})$.*

Proof. Let $p_1, \ldots, p_n$ be orthogonal Abelian projections in $\mathcal{M}$ with sum 1 and central supports 1. By 5.3.2 there exist symmetries $s_1, \ldots, s_n$ in $\mathcal{M}$ such that $s_i p_i s_i = p_1$. Let $e_{ij} = e_i s_i s_j e_j$. Then $\{e_{ij}: 1 \leq i, j \leq n\}$ is a set of $n \times n$ matrix units in $\mathcal{M}$ whose linear span is isomorphic to $M_n(\mathbb{C})$. By 5.2.17 $e_i \mathcal{M} e_i \cong \mathcal{Z} e_i$, so it follows as in the proof of 2.8.3 that $\mathcal{M} \cong M_n(\mathcal{Z})$.

7.4.6. Lemma. *Let $\mathcal{M}$ be a von Neumann algebra and $M = \mathcal{M}_{sa}$. Then M is reversible in every normal representation.*

Proof. By 5.3.10 it remains to consider the case when M is of type I_2. Let $\mathcal{Z}$ denote the centre of $\mathcal{M}$. By 7.4.5 $\mathcal{M} \cong M_2(\mathcal{Z})$, which we by tensor notation identify with $\mathcal{Z} \otimes M_2(\mathbb{C})$ under the map $(z_{ij}) \rightarrow \sum_{i,j=1}^{2} z_{ij} \otimes e_{ij}$, where $z_{ij} \in \mathcal{Z}$, and e_{ij} are the standard matrix units in $M_2(\mathbb{C})$. Under this identification we identify $\mathcal{Z}$ with $\mathcal{Z} \otimes 1$.

Let π be a normal representation of M on a complex Hilbert space. Then π extends by linearity to a Jordan $*$ homomorphism of $\mathcal{M}$, i.e. π preserves the special Jordan product. In order to prove the lemma we must show $\pi(\mathcal{M})$ is closed under symmetric products

$$\{x_1 \ldots x_m\} = x_1 \ldots x_m + x_m \ldots x_1.$$

By 6.2.5 the lemma is true for $M_2(\mathbb{C})$, hence $\pi(1 \otimes M_2(\mathbb{C}))$ is closed under symmetric products. Let $a_1, \ldots, a_n \in M_2(\mathbb{C})$ and $z_1, \ldots, z_n \in \mathscr{Z}$. Then

$$\pi(z_i \otimes a_i) = \pi(z_i \otimes 1)\pi(1 \otimes a_i),$$

and $\pi(z_i \otimes 1)$ belongs to the centre of $\pi(\mathcal{M})$. Since π is a homomorphism on $\mathscr{Z} \otimes 1$ we have

$$\begin{aligned} &\{\pi(z_1 \otimes a_1) \ldots \pi(z_n \otimes a_n)\} \\ &\quad = \{\pi(z_1 \otimes 1)\pi(1 \otimes a_1) \ldots \pi(z_n \otimes 1)\pi(1 \otimes a_n)\} \\ &\quad = \pi(z_1 \ldots z_n \otimes 1)\{\pi(1 \otimes a_1) \ldots \pi(1 \otimes a_n)\}, \end{aligned}$$

which belongs to $\pi(\mathcal{M})$. Since all operators in $\pi(M)$ can be written $\pi(\sum z_i \otimes a_i)$, and symmetric products of such operators are sums of symmetric products of elementary tensors $\pi(z_i \otimes a_i)$, $\pi(\mathcal{M})$ is closed under symmetric products. In particular $\pi(M)$ is reversible.

7.4.7. Theorem. *Let $\mathcal{M}$ be a von Neumann algebra without direct summands of type I_1. Let $M = \mathcal{M}_{sa}$. Then $W^*(M) \cong \mathcal{M} \oplus \mathcal{M}^0$, and the inclusion map of M into $W^*(M)$ is given by $a \to a \oplus a^0$.*

Proof. By 7.4.6 M is reversible in every normal representation. Therefore by 7.3.4 and 7.3.5 there are projections e and f in the centres Z of M and $\mathscr{Z}$ of $W^*(M)$ respectively such that $e + f + \Phi(f) = 1$, $\Phi(g) = g$ for each projection $g \in Ze$, and the theorem is true for $e^{\perp}M$. We shall complete the proof by showing $e = 0$. Suppose not, then by considering eM we may assume $e = 1$, hence that Φ leaves $\mathscr{Z}$ pointwise invariant. By 7.3.3 we therefore have $Z = \mathscr{Z}_{sa}$.

Let $\iota: M \to \mathcal{M}$ be the identity map. By 7.1.9 there is a normal $*$ homomorphism $\hat{\iota}: W^*(M) \to \mathcal{M}$ such that $\hat{\iota}\psi = \iota$, where ψ is the canonical imbedding of M into $W^*(M)$. Since the support projection of $\hat{\iota}$ is a central projection in $W^*(M)$ (4.3.7), it belongs to Z, hence is the identity. Therefore $\hat{\iota}$ is an isomorphism; in particular ψ is an isomorphism of $\mathcal{M} = M + iM$ onto $W^*(M)$.

Apply the preceding paragraph to the identity map $\iota_0: a \to a^0$ of M into $\mathcal{M}^0$. This map then induces an isomorphism $\hat{\iota}_0$ of $\mathcal{M}$ onto $\mathcal{M}^0$, via the identification of $\mathcal{M}$ with $W^*(M)$. But $\hat{\iota}_0$ extends the identity map, so it is thus also an anti-isomorphism of $\mathcal{M}$ onto $\mathcal{M}^0$. Therefore $\mathcal{M}$ is Abelian, contrary to assumption. Thus contradiction proves the theorem.

7.4.8. If $\mathcal{M}$ and $\mathcal{N}$ are C^* algebras a linear map $\phi: \mathcal{M} \to \mathcal{N}$ is a *Jordan homomorphism* if it preserves the special Jordan product and the $*$ operation. The next two results are applications of the previous results to the characterization of such maps.

7.4.9. Corollary. *Let $\mathcal{M}$ and $\mathcal{N}$ be von Neumann algebras and $\phi: \mathcal{M} \to \mathcal{N}$ a normal Jordan homomorphism. Then ϕ is the sum of a $*$ homomorphism and a $*$ antihomomorphism.*

Proof. We may assume $\mathcal{M}$ has no type I_1 part. Let $M = \mathcal{M}_{sa}$. By 7.4.7 $W^*(M) = \mathcal{M} \oplus \mathcal{M}^0$, and by 7.1.9 ϕ extends to a $*$ homomorphism $\hat{\phi}$ of $W^*(M)$ into $\mathcal{N}$. Composing $\hat{\phi}$ with the maps $a \to a \oplus 0$ and $a \to 0 \oplus a^0$ we get a $*$ homomorphism and a $*$ antihomomorphism with sum ϕ.

7.4.10. A von Neumann algebra is called a *factor* if its centre is the scalars, or equivalently, its self-adjoint part is a JBW factor.

7.4.11. Corollary. *A Jordan automorphism of a factor is either an automorphism or an antiautomorphism.*

Proof. Let ϕ be a Jordan automorphism of a factor $\mathcal{M}$. By 4.5.6 ϕ is normal, and by 7.4.9 $\phi = \phi_1 + \phi_2$ with ϕ_1 a homomorphism and ϕ_2 an antihomomorphism of $\mathcal{M}$ into itself. Since $\phi_i(1)$ $(i = 1, 2)$ are projections and $\phi_1(1) + \phi_2(1) = 1$, they are orthogonal. Since $\phi(\mathcal{M}) = \mathcal{M}$, $\phi_i(1)$ is a central projection for each i, hence $\phi_i(1)$ is 0 or 1, proving the corollary.

7.4.12. Remark. If $\mathcal{A}$ is a C^* algebra and $\phi: \mathcal{A} \to \mathcal{M}$ is a Jordan homomorphism of $\mathcal{A}$ into a von Neumann algebra $\mathcal{M}$ then ϕ is the sum of a homomorphism and an antihomomorphism. Indeed, by 4.5.7 ϕ has an extension to a normal Jordan homomorphism from $\mathcal{A}^{**}$ into $\mathcal{M}$, hence the assertion is immediate from 7.4.9.

We shall be interested in extending this result to Jordan homomorphisms into a second C^* algebra $\mathcal{B}$. However, we need more conditions for the result in this case, as can be seen from the following example. Let $\mathcal{A}$ be the subalgebra of $C([-1, 1], M_2(\mathbb{C}))$ consisting of all f such that $f(0)$ is a diagonal matrix. Let $\phi: \mathcal{A} \to \mathcal{A}$ be the Jordan homomorphism

$$\phi(f)(s) = \begin{cases} f(s), & s \geq 0 \\ f(s)^t, & s \leq 0 \end{cases}$$

where the t denotes the transpose. Then, if ϕ is a sum of a $*$ homomorphism ϕ_1 and a $*$ antihomomorphism ϕ_2, we must have

$$\phi_1(f)(s) \begin{cases} f(s), & s \geq 0 \\ 0, & s < 0 \end{cases}$$

but then $\phi_1(f)$ may be discontinuous, i.e. ϕ_1 does not map $\mathscr{A}$ into $\mathscr{A}$. The problem in this example is seen to arise from the presence of one-dimensional representations.

7.4.13. We define the *commutator ideal* $[\mathscr{A}, \mathscr{A}]$ of $\mathscr{A}$ to be the closed two-sided ideal generated by all commutators $[a, b] = ab - ba$, $a, b \in \mathscr{A}$. This is easily seen to be the smallest ideal J in $\mathscr{A}$ with the property that $\mathscr{A}/J$ is Abelian. Moreover, since any Abelian C^* algebra has a faithful family of one-dimensional representations, i.e. homomorphisms onto $\mathbb{C}$ (alias multiplicative states), $[\mathscr{A}, \mathscr{A}]$ is the intersection of the kernels of all one-dimensional representations of $\mathscr{A}$.

Similarly, we could consider the weakly closed ideal in a von Neumann algebra $\mathscr{M}$ generated by all commutators. But again this is seen to be the smallest weakly closed ideal J such that $\mathscr{M}/J$ is Abelian, so that if we decompose $\mathscr{M}$ according to types as

$$\mathscr{M} = \mathscr{M}_{I_1} \oplus \mathscr{M}_{I_2} \oplus \ldots \oplus \mathscr{M}_{I_\infty} \oplus \mathscr{M}_{II} \oplus \mathscr{M}_{III},$$

then this ideal is just the non-Abelian or non-type I_1 part, i.e. the sum of all components in the above direct sum except $\mathscr{M}_{I_1}$.

It is almost obvious from the above discussion that if $\mathscr{A}$ is a C^* algebra then the weak closure of $[\mathscr{A}, \mathscr{A}]$ in $\mathscr{A}^{**}$ is the non-Abelian part of $\mathscr{A}^{**}$.

7.4.14. Theorem. *Let $\mathscr{A}$ be a C^* algebra with no one-dimensional representations. Let $\mathscr{B}$ be another C^* algebra. Then any Jordan homomorphism of $\mathscr{A}$ into $\mathscr{B}$ is a sum of a $*$ homomorphism and a $*$ antihomomorphism.*

Proof. Since $\mathscr{A}$ has no one-dimensional representations, it follows from 7.4.13 that $[\mathscr{A}, \mathscr{A}] = \mathscr{A}$ and hence that $\mathscr{A}^{**}$ has no Abelian part. By 7.4.7, $W^*(\mathscr{A}^{**}_{sa}) = \mathscr{A}^{**} \oplus \mathscr{A}^{**0}$, and the inclusion $\psi: \mathscr{A}^{**}_{sa} \to W^*(\mathscr{A}^{**}_{sa})$ is given by $\psi(a) = a \oplus a^0$.

If $\phi: \mathscr{A} \to \mathscr{B}$ is a Jordan homomorphism, so is $\phi^{**}: \mathscr{A}^{**} \to \mathscr{B}^{**}$. By 7.1.9 and the above, ϕ^{**} defines a $*$ homomorphism $\tilde{\phi}: \mathscr{A}^{**} \oplus \mathscr{A}^{**0} \to \mathscr{B}^{**}$ such that $\tilde{\phi}\psi = \phi^{**}$. Let

$$\phi_1(a) = \tilde{\phi}(a \oplus 0), \qquad a \in \mathscr{A},$$

$$\phi_2(a) = \tilde{\phi}(0 \oplus a^0), \qquad a \in \mathscr{A}.$$

Then $\phi_i: \mathscr{A} \to \mathscr{B}^{**}$ and $\phi = \phi_1 \oplus \phi_2$. It remains to prove that ϕ_i maps $\mathscr{A}$ into $\mathscr{B}$. For this it is sufficient to show that $\mathscr{A} \oplus 0$ is contained in the C^* algebra generated by $\psi(\mathscr{A})$, because $\tilde{\phi}$ maps $\psi(\mathscr{A})$ into $\mathscr{B}$.

Let J be the set of all $a \in \mathscr{A}$ such that $a \oplus 0$ belongs to the C^* algebra generated by $\psi(\mathscr{A})$. Then J is an ideal, because $ab \oplus 0 = (a \oplus 0)(b \oplus b^0)$. It contains all commutators, because $[a, b]\oplus 0=(a\oplus a^0)(b\oplus b^0)-(ba)\oplus(ba)^0$. Hence $J\supseteq[\mathcal{A}, \mathcal{A}]$, or $J=\mathcal{A}$. Thus $\mathcal{A}\oplus 0$ is contained in the C^* algebra generated by $\psi(\mathcal{A})$.

7.4.15. Remark. Using 7.1.11, 7.4.7 and the above technique of proof it is not difficult to show that if $\mathcal{A}$ has no one-dimensional representations then $C_u^*(\mathcal{A}_{sa})\cong\mathcal{A}\oplus\mathcal{A}^0$. More generally,

$$C_u^*(\mathcal{A}_{sa})\cong\{a\oplus b^0\in\mathcal{A}\oplus\mathcal{A}^0 \colon b-a\in[\mathcal{A}, \mathcal{A}]\}.$$

Indeed, from 7.4.7 we get the analogous formula for a von Neumann algebra $\mathcal{M}$, where $\mathcal{M}_{na}$ denotes the non-Abelian part of $\mathcal{M}$:

$$W^*(\mathcal{M}_{sa})=\{a\oplus b^0\in\mathcal{M}\oplus\mathcal{M}^0 \colon b-a\in\mathcal{M}_{na}\}.$$

(Indeed, splitting $\mathcal{M}$ as $\mathcal{M}=\mathcal{M}_{I_1}\oplus\mathcal{M}_{na}$, the above formula is just a difficult way of writing $\mathcal{M}_{I_1}\oplus\mathcal{M}_{na}\oplus\mathcal{M}_{na}^0$.) From this, using 7.1.11 the proof is easy.

7.5. JBW factors of type I_∞

7.5.1. We have shown that JBW factors of type I_2 are spin factors, and were thus classified in 6.2.2. The JBW factors of type I_n, $3\leqslant n<\infty$, were classified in 5.3.8. Therefore, in order to obtain a complete classification of all JBW factors of type I it remains to classify all JBW factors of type I_∞. Furthermore, by 5.3.10 such JBW factors are isomorphic to reversible JW factors of type I_∞. We shall in the present section study these algebras in more detail. In order to do this we shall need to know that von Neumann algebras that are factors of type I are isomorphic to some $B(H)$, and then that their $*$ automorphisms are all inner.

7.5.2. Proposition. *Let $\mathcal{M}$ be a von Neumann algebra which is a factor of type* I. *Then there is a complex Hilbert space H such that $\mathcal{M}\cong B(H)$.*

Proof. By Zorn's lemma and 5.3.2 there is an orthogonal family $(p_\alpha)_{\alpha\in J}$ of minimal projections in the JBW algebra $\mathcal{M}_{sa}$ with sum 1. Fix $\alpha_0\in J$ and let for each $\alpha\in J$, s_α be a symmetry in $\mathcal{M}_{sa}$ such that $s_\alpha p_\alpha s_\alpha=p_{\alpha_0}$. Let $\mathcal{M}$ act on a Hilbert space K, let $K_0=p_{\alpha_0}(K)$, and let H be a complex Hilbert space of dimension equal to card J. Let $(\eta_\alpha)_{\alpha\in J}$ be an orthonormal basis for H. Define $u\colon K\to K_0\otimes H$ by

$$u\xi=\sum_\alpha p_{\alpha_0}s_\alpha\xi\otimes\eta_\alpha.$$

Then a straightforward computation shows that u is a unitary operator. If

$x \in \mathcal{M}$ and $\xi \in K_0$ then

$$uxu^{-1}(\xi \otimes \eta_\alpha) = uxs_\alpha\xi = \sum_\beta p_{\alpha_0}s_\beta xs_\alpha\xi \otimes \eta_\beta.$$

Let $w_{\alpha\beta}$ be the partial isometry $w_{\alpha\beta}\xi = (\xi, \eta_\beta)\eta_\alpha$ in $B(H)$. Then it follows that

$$uxu^{-1} = \sum_{\alpha,\beta} p_{\alpha_0}s_\beta xs_\alpha p_{\alpha_0} \otimes w_{\beta\alpha} = \sum_{\alpha,\beta} \omega_{\alpha\beta}(x)p_{\alpha_0} \otimes w_{\beta\alpha},$$

where $\omega_{\alpha\beta}(x) \in \mathbb{C}$ is defined in the obvious way since p_{α_0} is a minimal projection. Thus $uxu^{-1} \in \mathbb{C}_{K_0} \otimes B(H)$ where $\mathbb{C}_{K_0}$ is the scalar operators on K_0. The map $\phi: \mathcal{M} \to B(H)$ defined by

$$\phi(x) = \sum_{\alpha,\beta} \omega_{\alpha\beta}(x)w_{\beta\alpha},$$

is then an isomorphism such that $uxu^{-1} = 1_{K_0} \otimes \phi(x)$. Note that $w_{\alpha\beta} = \phi(p_\alpha s_\alpha s_\beta p_\beta) \in \phi(\mathcal{M})$. Therefore if e_F denotes the projection $e_F = \sum_{\beta \in F} w^*_{\alpha_0\beta}w_{\alpha_0\beta}$ for each finite subset $F \subset J$, e_F is the projection on the subspace spanned by $\{\eta_\beta : \beta \in F\}$, and $e_F B(H) e_F \subset \phi(\mathcal{M})$. Since u is unitary the map $x \to uxu^{-1}$ is a homomorphism of $\mathcal{M}$ onto $u\mathcal{M}u^{-1}$ in the ultraweak topology, hence $\phi(\mathcal{M})$ is ultraweakly closed. Let $\mathcal{F}$ denote the directed set of finite subsets of J ordered by inclusion. Then the net $(e_F)_{F \in \mathcal{F}}$ is increasing in $\phi(\mathcal{M})$ with least upper bound 1. Thus $e_F \to 1$ strongly in the sense of JBW algebras (4.1.3), hence $e_F x e_F \to x$ strongly for each $x \in B(H)$ (4.1.9). This shows $\phi(\mathcal{M}) = B(H)$, and the proof is complete.

7.5.3. Lemma. *Let H be a complex Hilbert space and α a $*$ automorphism of $B(H)$. Then there exists a unitary operator u such that $\alpha(x) = uxu^{-1}$ for all $x \in B(H)$.*

Proof. If p is a minimal projection in $B(H)$ then so is $\alpha(p)$. Therefore p is one-dimensional if and only if $\alpha(p)$ is one-dimensional. Fix such a p and choose unit vectors ξ, η in H such that $p\xi = \xi$ and $\alpha(p)\eta = \eta$. Define a linear operator on H by $ux\xi = \alpha(x)\eta$. Since $H = B(H)\xi = B(H)\eta$, u is everywhere defined and surjective. We show u is unitary by showing that it is an isometry. Since α is isometric and $\alpha(p)$ one-dimensional this follows from the following computation.

$$\begin{aligned}\|ux\xi\|^2 &= \|\alpha(x)\eta\|^2 = (\alpha(x)\eta, \alpha(x)\eta)\\ &= (\alpha(p)\alpha(x^*x)\alpha(p)\eta, \eta)\\ &= \|\alpha(px^*xp)\| = \|px^*xp\|\\ &= (x^*x\xi, \xi) = \|x\xi\|^2.\end{aligned}$$

Finally, if $x, y \in B(H)$ then

$$uxu^*\alpha(y)\eta = uxy\xi = \alpha(xy)\eta = \alpha(x)\alpha(y)\eta,$$

proving that $uxu^* = \alpha(x)$.

7.5.4. Let H be a complex Hilbert space with inner product $(\xi \mid \eta)$, $\xi, \eta \in H$. We make H into a real Hilbert space by defining a real inner product by

$$\langle \xi, \eta \rangle = \text{Re}\,(\xi, \eta).$$

A straightforward computation then shows that

$$(\xi, \eta) = \langle \xi, \eta \rangle - \mathrm{i}\langle \mathrm{i}\xi, \eta \rangle. \tag{7.1}$$

Let a be a bounded real linear operator on H. We define an adjoint a^+ of a with respect to the real Hilbert space structure by

$$\langle a\xi, \eta \rangle = \langle \xi, a^+\eta \rangle.$$

If i denotes multiplication on H by the scalar i then

$$\langle \xi, \mathrm{i}^+\eta \rangle = \langle \mathrm{i}\xi, \eta \rangle = \text{Re}\,(\mathrm{i}\xi, \eta) = -\text{Re}\,(\xi, \mathrm{i}\eta) = -\langle \xi, \mathrm{i}\eta \rangle,$$

hence $\mathrm{i}^+ = -\mathrm{i} = \mathrm{i}^*$, where $*$ denotes the usual adjoint. If a is a complex linear operator on H then $\mathrm{i}a = a\mathrm{i}$, so that $\mathrm{i}a^+ = -\mathrm{i}^+a^+ = -(a\mathrm{i})^+ = -(\mathrm{i}a)^+ = -a^+\mathrm{i}^+ = a^+\mathrm{i}$, so that a^+ is complex linear. Since clearly $\langle \xi, a^*\eta \rangle = \langle \xi, a^+\eta \rangle$ it therefore follows from (7.1) that $a^* = a^+$. Thus we may write a^* rather than a^+ in the general case without fear of confusion.

Suppose a is a conjugate linear operator on H, i.e. a is real linear and $a\lambda\xi = \bar{\lambda}a\xi$, $\lambda \in \mathbb{C}$. Then similarly a^* is conjugate linear and from (7.1)

$$(a\xi, \eta) = (a^*\eta, \xi). \tag{7.2}$$

7.5.5. An *antiunitary* operator on H is a conjugate linear isometry v of H onto itself. A *conjugation* on H is an antiunitary operator J on H such that $J^2 = 1$. A *unit quaternion* is an antiunitary operator j such that $j^2 = -1$.

Note that with v, J, j as above then $v^*v = vv^* = 1$, $J^* = J$, $j^* = -j$. Let H^J denote the fixed point set of J in H. Then H^J is a real Hilbert space in the inner product (7.1) on H. Indeed, H^J is clearly a real subspace of H, and if $\xi, \eta \in H^J$ then by (7.2)

$$(\xi, \eta) = (J\xi, \eta) = (J\eta, \xi) = (\eta, \xi)$$

is real. It follows that $H = H^J \oplus \mathrm{i}H^J$ is the complexification of H^J, and $J(\xi + \mathrm{i}\eta) = \xi - \mathrm{i}\eta$ when $\xi, \eta \in H^J$.

If j is a unit quaternion and i is multiplication by the scalar i then $\{1, \mathrm{i}, j, k = \mathrm{i}j\}$ is a basis for the quaternions (cf. proof of 2.2.6). Indeed we have $ij = k = -ji$, $jk = i = -kj$, $ki = j = -ik$, and $i^2 = j^2 = k^2 = -1$. We can define a quaternionian inner product on H by

$$(\xi, \eta)_{\mathbb{H}} = \langle \xi, \eta \rangle - i\langle i\xi, \eta \rangle - j\langle j\xi, \eta \rangle - k\langle k\xi, \eta \rangle.$$

Then we have after trivial but rather tedious computations

$$(\xi, \eta)_{\mathbb{H}} = \overline{(\eta, \xi)_{\mathbb{H}}},$$

$$(\lambda\xi, \eta)_{\mathbb{H}} = \lambda(\xi, \eta)_{\mathbb{H}}, \qquad (\xi, \lambda\eta)_{\mathbb{H}} = (\xi, \eta)\bar{\lambda}, \qquad \lambda \in \mathbb{H}.$$

7.5.6. Lemma. *If J and J' (resp. j and j') are two conjugations (resp. unit quaternions) on H then there exists a unitary operator u on H such that $J' = uJu^*$ (resp. $j' = uju^*$). Furthermore, there always exists a conjugation on H, and there exists a unit quaternion on H if and only if* $\dim H$ *is either an even number or infinite.*

Proof. If J and J' are two conjugations let (ξ_α) and (η_α) be orthonormal bases for H^J and $H^{J'}$ respectively. Then they are both orthonormal bases for H, so u defined by $u\xi_\alpha = \eta_\alpha$ is the desired unitary operator. The existence of a conjugation is proved by choosing an orthonormal basis (ξ_α) for H and defining $J(\sum \lambda_\alpha \xi_\alpha) = \sum \bar{\lambda}_\alpha \xi_\alpha$.

Suppose j is a unit quaternion. If $\xi \in H$ then by (7.2) $(j\xi, \xi) = (j^*\xi, \xi) = -(j\xi, \xi) = 0$, so $j\xi \perp \xi$. Let (ξ_α) be a maximal orthonormal family of vectors in H such that $\xi_\alpha \perp j\xi_\beta$ for all α, β. Then $\{\xi_\alpha, j\xi_\alpha\}$ is an orthonormal basis for H. Indeed, if not then there is a unit vector η orthogonal to all $\xi_\alpha, j\xi_\alpha$. But then $(j\eta, \xi_\alpha) = (j^*\xi_\alpha, \eta) = -(j\xi_\alpha, \eta) = 0$, and $(j\eta, j\xi_\alpha) = (\eta, \xi_\alpha) = 0$ for all α, so that both η and $j\eta$ are orthogonal to all $\xi_\alpha, j\xi_\alpha$, contradicting the maximality of (ξ_α). This proves the assertion. In particular $\dim H$ is an even number or infinite.

Let j' be another unit quaternion and choose a corresponding family (η_α) for j'. Define a unitary operator u on H by $u\xi_\alpha = \eta_\alpha$, $uj\xi_\alpha = j'\eta_\alpha$. Then $uju^* = j'$.

Finally, suppose $\dim H$ is an even number or infinite. Let $\{\xi_\alpha, \eta_\alpha\}$ be an orthonormal basis for H. Define $j\xi_\alpha = \eta_\alpha$ and $j\eta_\alpha = -\xi_\alpha$, and extend j to be a conjugate linear operator. Then j becomes a unit quaternion. The proof is complete.

7.5.7. In general if v is an antiunitary operator on H then we can define a $*$ antiautomorphism on $B(H)$ by

$$\beta(a) = va^*v^*. \tag{7.3}$$

Then β^2 is the $*$ automorphism implemented by v^2. In particular $\beta^2 = \iota$ if and only if $v^2 = \lambda 1$ for a complex number λ of modulus 1. Since v commutes with v^2, v commutes with λ, and thus λ is real since v is conjugate linear. Therefore $\lambda = \pm 1$, and we have shown $\beta^2 = \iota$ if and only if $v^2 = \pm 1$, or rather v is either a conjugation or a unit quaternion.

7.5.8. An antiautomorphism β of $B(H)$ of the form (7.3) is called a *real flip* if $v^2 = +1$ and a *quaternionian flip* if $v^2 = -1$.

Let J be a conjugation on H, and let α denote the real flip $\alpha(a) = Ja^*J$. If β is any $*$ antiautomorphism of $B(H)$ then $\beta\alpha$ is a $*$ automorphism, hence is the form $\beta\alpha = \operatorname{Ad} u$, where u is the unitary operator and $\operatorname{Ad} u(a) = uau^*$ (7.5.3). In particular $\beta(a) = va^*v^*$ where $v = uJ$.

7.5.9. Lemma. *Let α and β be $*$ antiautomorphisms of order 2 of $B(H)$. Then there exists a $*$ automorphism γ of $B(H)$ such that $\beta = \gamma\alpha\gamma^{-1}$ if and only if either both α and β are real flips or both are quaternionian flips.*

Proof. If for example $\alpha(a) = Ja^*J$ and $\beta(a) = J'a^*J'$ with J and J' conjugations let by 7.5.6 u be a unitary operator such that $J' = uJu^*$. Then $\beta = \operatorname{Ad} u\alpha(\operatorname{Ad} u)^{-1}$, and similarly if α and β are quaternionian flips.

Conversely suppose α is a real flip and $\beta = \gamma\alpha\gamma'$. Then γ, being an automorphism of $B(H)$, is of the form $\gamma = \operatorname{Ad} u$ (7.5.3). Let $J' = uJu^*$. Then $\beta(a) = J'a^*J'$, so is a real flip. The proof in the quaternionian case is similar.

7.5.10. Let α be a real flip on $B(H)$ for the Hilbert space H, say $\alpha(x) = Jx^*J$, where J is a conjugation of H. If $x \in B(H)^{\alpha}_{sa}$ then $x = JxJ$, or $Jx = xJ$. This means that x leaves H^J invariant, and since $H = H^J \oplus iH^J$, x is the complexification of its restriction to H^J. Therefore restriction to H^J is an isomorphism of $B(H)^{\alpha}_{sa}$ onto the Jordan algebra of all symmetric bounded linear operators on the real Hilbert space H^J.

Similarly, let β be a quaternionian flip on $B(H)$, say $\beta(x) = -jx^*j$, where j is a unit quaternion. As noted in 7.5.5 j induces a structure of quaternionic Hilbert space on H. We write $H_{\mathbb{H}}$ for H with this added structure. If $x \in B(H)^{\beta}_{sa}$ then $x = -jxj$ or $jx = xj$. From this we also get $kx = xk$ where $k = ij$, so this means that x is $\mathbb{H}$-linear. In other words, $B(H)^{\beta}_{sa}$ is the Jordan algebra of $\mathbb{H}$-linear self-adjoint bounded operators on the quaternionic Hilbert space $H_{\mathbb{H}}$.

7.5.11. Theorem. *Let M be a* JBW *factor of type* I_n, $4 \leq n \leq \infty$. *Then there exists a complex Hilbert space H such that M is isomorphic to one of the following three* JW *factors:*

(i) $M \cong B(H)_{sa}$.
(ii) $M \cong B(H)^{\alpha}_{sa}$ *with α a real flip on $B(H)$.*
(iii) $M \cong B(H)^{\beta}_{sa}$ *with β a quaternionian flip on $B(H)$.*

Proof. As remarked in 7.5.1 we may assume M is a JW algebra acting on a complex Hilbert space. If M is isomorphic to the self-adjoint part of a von Neumann algebra we have (i) (7.5.2). Let Φ be the canonical antiautomorphism of $W^*(M)$ (7.1.9). If M is not isomorphic to the self-adjoint part of a von Neumann algebra then by 7.3.5 the centre $\mathscr{Z}$ of $W^*(M)$ is pointwise Φ-invariant, hence $\mathscr{Z}$ equals the centre of M by 7.3.3. Thus $W^*(M)$ is a factor. By 7.4.2 $W^*(M)$ is of type I, hence is by 7.5.2 isomorphic to $B(H)$ for some complex Hilbert space H. Let γ be the antiautomorphism of $B(H)$ corresponding to Φ under this isomorphism. By 7.3.3 M corresponds to $B(H)_{sa}^{\Phi}$ under this isomorphism. An application of 7.5.9 completes the proof.

7.5.12. Remark. Let M be an irreducible JW algebra acting on a Hilbert space H, i.e. the identity is the only nonzero projection in $B(H)$ commuting with all elements in M. Then M is a JW factor of type I, hence is either a spin factor or a factor of the types described in 7.5.11. Indeed, it is an easy consequence of the double commutant theorem for von Neumann algebras [6, Ch. 1, §3, Thm 2] that the only irreducible von Neumann algebra on H is $B(H)$. Thus $B(H)$ is the von Neumann algebra generated by M. Since $B(H)$ is of type I, so is M by 7.4.3.

7.5.13. Lemma. *Let $\mathcal{M}$ be a von Neumann algebra. Suppose there is a normal state which is pure on $\mathcal{M}$. Then $\mathcal{M}$ has a direct summand of type* I.

Proof. Let ρ be a pure normal state on $\mathcal{M}$, and let e be the central support of ρ, i.e. the smallest central projection in $\mathcal{M}$ such that $\rho(e)=1$. Then e is minimal in the centre. Indeed, suppose $0<f\leqslant e$ is a central projection, and let $\sigma\in\mathcal{M}_*^+$ be defined by $\sigma(x)=\rho(fx)$. Then $0\leqslant\sigma\leqslant\rho$, so by 1.2.1 $\sigma=\rho(f)\rho$. In particular, $\rho(f)=\rho(f^2)=\rho(f)^2$, so $\rho(f)=0$ or 1. Therefore $f=e$ by definition of e.

By the above we may assume $\mathcal{M}$ is a factor. Otherwise replace $\mathcal{M}$ by $e\mathcal{M}$. Let $\rho(x)=(\pi(x)\xi_\rho,\xi_\rho)$ be the GNS representation of ρ (1.3.10), where π is a representation of $\mathcal{M}$ on a Hilbert space H_ρ. Let (a_α) be a bounded decreasing net in $\mathcal{M}^+$ such that $a_\alpha\searrow 0$. Then for all $y\in\mathcal{M}$, $y^*a_\alpha y\searrow 0$, so that

$$(\pi(a_\alpha)y\xi_\rho, y\xi_\rho)=(\pi(y^*a_\alpha y)\xi_\rho,\xi_\rho)=\rho(y^*a_\alpha y)\searrow 0.$$

Since vectors of the form $y\xi_\rho$ are dense in H_ρ, and the net (a_α) is bounded, $(\pi(a_\alpha)\xi,\xi)\searrow 0$ for all $\xi\in H_\rho$. It is then easy to see $\omega(\pi(a_\alpha))\searrow 0$ for all $\omega\in B(H)_*^+$ (see 1.4.1), hence that $\pi(a_\alpha)\searrow 0$. Thus π is normal, since if $b_\alpha\nearrow b$ then $b-b_\alpha\searrow 0$. By 4.5.11 $\pi(\mathcal{M}_{sa})$ is a JW algebra, hence $\pi(\mathcal{M})$ is a von Neumann algebra by 1.4.6 and 4.5.10. Furthermore, by normality of π the kernel of π is of the form $e\mathcal{M}$ with e a central projection (4.3.7). Since $\mathcal{M}$ is a factor $e=0$, so π is an isomorphism. Finally by 1.4.8 $\pi(\mathcal{M})'=\mathbb{C}1$, hence as pointed out in 7.5.12, $\pi(\mathcal{M})=B(H_\rho)$, which is of type I. Thus $\mathcal{M}$, being isomorphic to $\pi(\mathcal{M})$, is of type I.

7.5.14. Proposition. *Let M be a* JBW *factor. Then M is of type* I *if and only if there exists a normal state which is pure on M.*

Proof. If M is of type I let e be a minimal projection in M. By 4.1.13 there is a normal state ω on M with support e. Since $M_e=\mathbb{R}e$ and each state ρ on M majorized by a multiple of ω has support e, $\rho=\omega$, and ω is pure by 1.2.1.

Conversely suppose ω is a pure normal state on M. In order to show M is of type I we may by 7.3.3 assume $M=W^*(M)^{\Phi}_{\mathrm{sa}}$, where Φ is the canonical antiautomorphism of the universal von Neumann algebra $W^*(M)$. Let by 1.2.4 ρ be a pure state extension of ω to $W^*(M)$ (i.e. $\rho \mid W^*(M)_{\mathrm{sa}}$ is pure). We show ρ is normal. By 7.3.2 $W^*(M)=R+iR$, where $R=\{x\in W^*(M)\colon \Phi(x)=x^*\}$. Suppose $a=b+ic\in W^*(M)^+$, $b,c\in R$. Then $b=b^*\in M$, and $c=-c^*$. Since $\Phi(a)=b^*+ic^*=b-ic\in W^*(M)^+$, $b=\frac{1}{2}(a+\Phi(a))\in M^+$ and $ic=\frac{1}{2}(a-\Phi(a))$. In particular $-b+ic=-\Phi(a)\leqslant 2ic\leqslant a=b+ic$, so $-b\leqslant ic\leqslant b$. Let now $(a_\alpha=b_\alpha+ic_\alpha)$ be a monotone decreasing net in $W^*(M)^+$ such that $a_\alpha\searrow 0$. Since Φ is in particular an order automorphism of $W^*(M)_{\mathrm{sa}}$, Φ is normal. Thus $\Phi(a_\alpha)\searrow 0$, whence $b_\alpha\searrow 0$, and so $\rho(b_\alpha)=\omega(b_\alpha)\searrow 0$. But then $|\rho(ic_\alpha)|\leqslant\rho(b_\alpha)$ converges to 0, and so $\rho(a_\alpha)\searrow 0$. Thus ρ is normal as asserted. By 7.5.13 $W^*(M)$ has a direct summand of type I. Since M is a JBW factor it is then of type I by 7.4.2.

7.6. Modularity in JBW algebras

7.6.1. As was indicated in Section 5.1 modularity in JBW algebras is closely related to finiteness in von Neumann algebras. Recall that in von Neumann algebras equivalence of projections is defined differently from equivalence in JBW algebras. In a von Neumann algebra two projections p and q are equivalent if there is a partial isometry v in the algebra such that $v^*v=p$ and $vv^*=q$. Thus it is possible that the identity 1 is equivalent to a proper subprojection. This is impossible in JBW algebras. Therefore we cannot as in von Neumann algebras say a JBW algebra is finite if the identity is never equivalent to a proper subprojection. It turns out, however, that modularity is the correct concept, and it is possible to extend most of the theory of finite von Neumann algebras to modular JBW algebras. We shall in the present section show two such results.

7.6.2. Lemma. *Let H be a separable infinite-dimensional real Hilbert space. Then we have:*

(i) *There exists a bounded symmetric operator a on H such that $a(H)$ is dense in H but is not a closed subspace.*

(ii) *There exist two closed subspaces P and Q of H such that $P+Q=\{\xi+\eta: \xi\in P, \eta\in Q\}$ is dense in H but not closed, and $P\cap Q=\{0\}$.*

Proof. (i) Let $(\xi_n)_{n\in\mathbb{N}}$ be an orthonormal basis for H. Let a be the unique symmetric operator of norm 1 on H such that $a\xi_n=(1/n)\xi_n$. Since $\xi_n\in a(H)$ for all n, $a(H)$ is dense in H. Let $\xi=\sum_{n=1}^{\infty}(1/n)\xi_n\in H$. Then $\xi\notin a(H)$, for if $\xi=a\eta$ with $\eta=\sum\lambda_n\xi_n\in H$, then $\xi=\sum(1/n)\lambda_n\xi_n$, which implies $\lambda_n=1$ for all n, contradicting the fact that $\sum\lambda_n^2<\infty$.

(ii) We can choose a subspace K of H such that H is identified with $K\oplus K$. Let $P=K\oplus 0$, and let by (i) a be a bounded symmetric operator on K such that $a(K)$ is dense in K but not closed. Since a is continuous its graph $Q=\{\xi\oplus a\xi: \xi\in K\}$ is a closed subspace of H. Then $P+Q=K\oplus a(K)$ is dense but not closed. Clearly $P\cap Q=\{0\}$.

7.6.3. Theorem. *Let M be a* JBW *algebra with projection lattice P. If $p\in P$ the following four conditions are equivalent:*

(i) *p is modular.*

(ii) *If $r, q\in P$, $r\leq q\leq p$, and r and q are perspective in $\{pMp\}$, then $r=q$.*

(iii) *There is no infinite sequence $(p_n)_{n\in\mathbb{N}}$ of pairwise orthogonal projections in M such that $p_n\leq p$, $p_m\sim p_n$ for all m, n.*

(iv) *$\{pMp\}$ contains no copy of $B(H)_{\mathrm{sa}}^{\alpha}$ for H a separable infinite-dimensional complex Hilbert space and α a real flip.*

Proof. The equivalence (i)$\Leftrightarrow$(ii) holds in any orthomodular lattice, see 5.1.3. We shall show (ii)$\Rightarrow$(iv)$\Rightarrow$(iii)$\Rightarrow$(ii).

(ii)$\Rightarrow$(iv) Let H be a separable infinite-dimensional complex Hilbert space and α a real flip. We shall prove that the identity in $B(H)_{\mathrm{sa}}^{\alpha}$ does not satisfy (ii). By 7.5.10 and 7.5.11 $B(H)_{\mathrm{sa}}^{\alpha}$ is isomorphic to the JBW algebra $B(K)_{\mathrm{s}}$ of bounded symmetric operators on a separable infinite-dimensional real Hilbert space K. By 7.6.2 there exist two projections r and r' in $B(K)_{\mathrm{s}}$ such that $r\wedge r'=0$ and $r\vee r'=1$, but $r(K)+r'(K)$ is not closed. Pick any vector $\xi\in K$ not lying in the sum, and let q be the projection on the closed subspace generated by $r(K)$ and ξ. Then $r<q$, and r, q have a common complement r', contradicting (ii).

(iv)$\Rightarrow$(iii) Assume $(p_n)_{n\in\mathbb{N}}$ is a sequence as described in (iii). We shall construct a copy of $B(H)_{\mathrm{sa}}^{\alpha}$ in $\{pMp\}$, and thus contradict (iv). For each $n\in\mathbb{N}$ let s_n be a partial symmetry such that $s_n^2=p_1+p_n$, $\{s_np_1s_n\}=p_n$, and let $s_1=p_1$. Define $s_{ii}=p_i$, $s_{ij}=2\{s_ip_1s_j\}$ if $i\neq j$. We have the following multiplication table for the s_{ij}:

$$\begin{aligned} s_{ij} \circ s_{kl} &= 0 && \text{if } \{i, j\} \cap \{k, l\} = \varnothing, \\ s_{ii}^2 &= s_{ii}, \\ s_{ii} \circ s_{ij} &= \tfrac{1}{2} s_{ij} && \text{if } i \neq j, \\ s_{ij}^2 &= s_{ii} + s_{jj} && \text{if } i \neq j, \\ s_{ij} \circ s_{jk} &= s_{ik} && \text{if } i, j, k \text{ distinct.} \end{aligned}$$

Let A be the linear span of all the s_{ij}. Then from the above multiplication table A is a Jordan subalgebra of M, hence the weak closure N of A is a JBW subalgebra. Since $\sum_{n=1}^{\infty} p_n$ is the identity in N and all p_n are minimal equivalent projections in N, N is a JBW factor of type I_∞. Since $\{(p_i + p_j)N(p_i + p_j)\}$ is the linear span of s_{ii}, s_{jj} and s_{ij} for all pairs $i \neq j$, and so is of real dimension 3, it follows from 7.5.11 that $N \cong B(H)_{\mathrm{sa}}^{\alpha}$ with α a real flip. Otherwise the dimension would have been 4 (in the complex case) or 6 (in the quaternionian case).

(iii)$\Rightarrow$(ii) Suppose $r < q$ are perspective projections in $\{pMp\}$. By 5.2.3 $r \underset{2}{\sim} q$ in $\{pMp\}$, say $\alpha = U_s U_t$ is an inner automorphism of $\{pMp\}$ mapping q to r. Let $p_1 = q - r$, $p_n = \alpha^{n-1}(p_1)$ for $n \geqslant 2$. Since $p_n = \alpha^{n-1}(q) - \alpha^n(q)$, $\sum_{n=1}^{m} p_n = q - \alpha^m(q)$ is a projection for each $m \in \mathbb{N}$. Thus all the p_n are pairwise orthogonal projections such that $p_1 \underset{2}{\sim} p_2 \underset{2}{\sim} p_3 \underset{2}{\sim} \dots$. A repeated application of 5.2.6 yields $p_n \underset{1}{\sim} p_m$ for all n, m. Since $p_n \leqslant p$ for all n, we have thus violated (iii).

7.6.4. Theorem. *Let M be a* JBW *algebra with projection lattice P. If p and q in P are modular then so is $p \vee q$.*

Proof. By 5.2.3 $p \vee q - p \underset{1}{\sim} q - p \wedge q \leqslant q$, so $p \vee q - p$ is modular. We may thus replace q by $p \vee q - p$ and assume $p \perp q$. Considering $\{p \vee q M p \vee q\}$ instead of M we may also assume $p \vee q = p + q = 1$. If 1 is not modular we can by 7.6.3 find an orthogonal sequence (p_n) in P such that $p_m \underset{1}{\sim} p_n$ for all $m, n \in \mathbb{N}$. Let $p' = \sum_{k=1}^{\infty} p_{2k}$, $q' = \sum_{k=1}^{\infty} p_{2k+1}$. By the comparison theorem (5.2.13) there is a central projection e in M such that $ep' \underset{1}{\lesssim} ep$ and $e^{\perp}p \underset{1}{\lesssim} e^{\perp}p'$. Therefore ep' is modular, whence by 7.6.3 $ep_{2k} = 0$ for all k, and so $ep' = 0$. If s is a symmetry such that $U_s e^{\perp} p \leqslant e^{\perp} p'$ then $U_s e^{\perp}(1-p) \geqslant e^{\perp}(1-p')$, hence we have

$$e^{\perp} q' \leqslant e^{\perp}(1-p') \underset{1}{\lesssim} e^{\perp}(1-p) = e^{\perp} q.$$

By the above argument $e^{\perp} q' = 0$, so that $ep_1 = p_1 \underset{1}{\sim} p_2 = e^{\perp} p_2$, which is impossible unless $p_1 = p_2 = 0$.

7.7. Comments

Universal C^* algebras were first introduced by Alfsen, Hanche-Olsen and Shultz [15] and then studied by Hanche-Olsen [57], who realized that they could be used to simplify many arguments in the theory. The main results of Section 7.2 are due to Alfsen, Shultz and Størmer; 7.2.3 was proved in Alfsen *et al.* [19] and 7.2.7 in Shultz [91], but the proofs were somewhat different. For further information on the exceptional ideal the reader is referred to the paper of Behncke and Bös [22].

Theorem 7.4.3 is due to Størmer [103, 105]. This result has recently been extended to types II_1, II_∞, III by Ajupov [13].

The results on Jordan homomorphisms in Section 7.4 were first proved for matrix algebras by Jacobson and Rickart [63]. Then they were extended to C^* algebras by Kadison [70] and Størmer [102]. The version presented in 7.4.14 first appeared in Hanche-Olsen [57] together with 7.4.15.

The early results in Section 7.5 are quite standard. The main results 7.5.11 and 7.5.14 are due to Størmer [103, 105].

Section 7.6 is inspired by the work of Topping [110] and the analogous results in von Neumann algebras. By modifying the proof for von Neumann algebras the reader should have no difficulty in showing that a JBW algebra with modular projection lattice has a normal trace, i.e. a normal state τ such that $\tau(\{sxs\}) = \tau(x)$ for all x and all symmetries in the JBW algebra. For a discussion of the many possible other definitions of trace the reader is referred to Pedersen and Størmer [83].

Bibliography

Books

1 E. M. Alfsen: *Compact Convex Sets and Boundary Integrals*, Ergebnisse der Math. 57, Springer-Verlag, Berlin, 1971.

2 L. Asimov and A. J. Ellis: *Convexity Theory and its Applications in Functional Analysis*, Academic Press, London, 1980.

3 O. Bratteli and D. W. Robinson: *Operator Algebras and Quantum Statistical Mechanics*, Springer-Verlag, New York, 1979.

4 H. Braun and M. Koecher: *Jordan-algebren*. Grundlehren der Math. Wissenschaften 128, Springer-Verlag, Berlin, Heidelberg and New York, 1966.

5 J. Dixmier: *Les C*-Algebres et leurs Representations*, Gauthier-Villars, Paris, 1969.

6 J. Dixmier: *Les Algebres d'Operateurs dans l'Espace Hilbertien*, Gauthier-Villars, Paris, 1969.

7 N. Dunford and J. T. Schwartz: *Linear Operators*, Interscience, New York, 1958.

8 G. G. Emch: *Algebraic Methods in Statistical Mechanics and Quantum Field Theory*, Wiley-Interscience, New York, 1972.

9 N. Jacobson: *Structure and Representations of Jordan Algebras*, Am. Math. Soc. Colloq. Publ. 39, Providence, RI, 1968.

10 O. Loos: *Bounded Symmetric Domains and Jordan Pairs*, Math. Lectures, Univ. of California, Irvine, 1977.

11 R. D. Schafer: *An Introduction to Nonassociative Algebras*, Academic Press, New York, 1966.

12 S. Strătilă and L. Zsidó: *Lectures on von Neumann Algebras*, Abacus Press, Tunbridge Wells, Kent, 1979.

Papers

13 S. A. Ajupov: Extension of traces and type criterions for Jordan algebras of self-adjoint operators, *Math. Z.*, **181** (1982), 253–68.

14 A. A. Albert: On a certain algebra of quantum mechanics, *Annals Math.*, **35** (1934), 65–73.

15 E. M. Alfsen, H. Hanche-Olsen and F. W. Shultz: State spaces of C^*-algebras, *Acta Math.*, **144** (1980), 267–305.
16 E. M. Alfsen and F. W. Shultz: Non-commutative spectral theory for affine function spaces on convex sets, *Mem. Am. Math. Soc.*, **172** (1976).
17 E. M. Alfsen and F. W. Shultz: On non-commutative spectral theory and Jordan algebras, *Proc. Lond. Math. Soc.* (3), **38** (1979), 497–516.
18 E. M. Alfsen and F. W. Shultz: State spaces of Jordan algebras, *Acta Math.*, **140** (1978), 155–90.
19 E. M. Alfsen, F. W. Shultz and E. Størmer: A Gelfand–Neumark theorem for Jordan algebras, *Adv. Math.*, **28** (1978), 11–56.
20 K. Alvermann: Real and complex noncommutative Jordan Banach factors, Preprint, 1982.
21 H. Behncke: Hermitian Jordan Banach algebras, *J. Lond. Math. Soc.* (2), **20** (1979), 327–33.
22 H. Behncke and W. Bös: JB algebras with an exceptional ideal, *Math. Scand.*, **42** (1978), 306–12.
23 J. Bellissard and B. Iochum: Homogeneous self-dual cones, versus Jordan algebras. The theory revisited, *Ann. Inst. Fourier*, **28,** 1 (1978), 27–67.
24 J. Bellissard and B. Iochum: L'algebre de Jordan d'un cone autopolaire facialement homogene, *C. R. Acad. Sci. Paris A*, **288** (1979), 229–32.
25 J. Bellissard and B. Iochum: Homogeneous self-dual cones and Jordan algebras, in L. Streit (ed.), *Quantum Fields—Algebras, Processes*, Springer-Verlag, Wien and New York, 1980, pp. 152–65.
26 J. Bellissard and B. Iochum: Spectral theory for facially homogeneous symmetric self-dual cones, *Math. Scand.*, **45** (1979), 118–26.
27 J. Bellissard and B. Iochum: Order structure and Jordan Banach algebras, *Proc. Symp. Pure Math.*, **38,** 2 (1982), 297–9.
28 F. F. Bonsall: Jordan algebras spanned by hermitian elements of a Banach algebra, *Math. Proc. Camb. Phil. Soc.*, **81** (1977), 3–13.
29 F. F. Bonsall: Jordan subalgebras of Banach algebras, *Proc. Edinb. Math. Soc.*, **21** (1978), 103–10.
30 F. F. Bonsall and P. Rosenthal: Certain Jordan operator algebras and double commutant theorems, *J. Funct. Anal.*, **21** (1976), 155–86.
31 H. N. Boyadjiev and M. A. Youngson: Alternators on Banach Jordan algebras, *C. R. Acad. Bulg. Sci.*, **33** (1980), 1589–90.
32 R. Braun, W. Kaup and H. Upmeier: A holomorphic characterization of Jordan C^*-algebras, *Math. Z.*, **161** (1978), 277–90.
33 R. Braun, W. Kaup and H. Upmeier: On the automorphisms of circular and Reinhardt domains in complex Banach spaces, *Man. Math.*, **25** (1978), 97–133.
34 Cho-Ho Chu: On the Radon–Nikodym property in Jordan algebras, *Operator Algebras and Group Representations*, Monographs and Studies in Mathematics, **17,** Pitman (1984), 65–70.
35 P. Civin and B. Yood: Lie and Jordan structures in Banach algebras, *Pacific J. Math.*, **15** (1965), 775–97.
36 C. M. Edwards: Ideal theory in JB-algebras, *J. Lond. Math. Soc.* (2), **16** (1977), 507–13.

37 C. M. Edwards: On the facial structure of a JB-algebra, *J. Lond. Math. Soc.* (2), **19** (1979), 335–44.

38 C. M. Edwards: Multipliers of JB-algebras, *Math. Ann.*, **249** (1980), 265–72.

39 C. M. Edwards: On the centres of hereditary JBW-subalgebras of a JBW-algebra, *Math. Proc. Camb. Phil. Soc.*, **85** (1979), 317–24.

40 C. M. Edwards: On Jordan W^*-algebras, *Bull. Soc. Math.*, **104** (1980), 393–403.

41 E. Effros and E. Størmer: Jordan algebras of self-adjoint operators, *Trans. Am. Math. Soc.*, **127** (1967), 313–16.

42 E. Effros and E. Størmer: Positive projections and Jordan structure in operator algebras, *Math. Scand.*, **45** (1979), 127–38.

43 G. G. Emch and W. P. C. King: Faithful normal states on JBW-algebras, *Proc. Symp. Pure Math.*, **38,** 2 (1982), 305–7.

44 Y. Friedman and B. Russo: Contractive projections on operator triple systems, *Math. Scand.*, **52** (1983), 279–311.

45 Y. Friedman and B. Russo: Solution of the contractive projection problem, Preprint, 1983.

46 Y. Friedman and B. Russo: Operator algebras without order: Complete solution to the contractive projection problem, Preprint, 1983.

47 Y. Friedman and R. Russo: Conditional expectation without order, Preprint, 1983.

48 T. Giordano: Antiautomorphismes involutifs des facteurs injectifs, *C. R. Acad. Sci. Paris A*, **291** (1980), 583–5.

49 T. Giordano: Antiautomorphismes involutifs des facteurs injectifs II, Preprint, 1982.

50 T. Giordano: Antiautomorphismes involutifs des facteurs de von Neumann injectifs, These, Neuchatel, 1981.

51 T. Giordano and V. Jones: Antiautomorphismes involutifs du facteur hyperfini de type II_1, *C. R. Acad. Sci. Paris A*, **290** (1980), 29–31.

52 C. M. Glennie: Some identities valid in special Jordan algebras but not valid in all Jordan algebras, *Pacific J. Math.*, **16** (1966), 47–59.

53 U. Haagerup and H. Hanche-Olsen: Tomita–Takesaki theory for Jordan algebras, *J. Operator Theory* (to appear).

54 H. Hanche-Olsen: Split faces and ideal structure of operator algebras, *Math. Scand.*, **48** (1981), 137–44.

55 H. Hanche-Olsen: A note on the bidual of a JB-algebra, *Math. Z.*, **175** (1980), 29–31.

56 H. Hanche-Olsen: A Tomita–Takesaki theory for JBW-algebras, *Proc. Symp. Pure Math.*, **38,** 2 (1982), 301–3.

57 H. Hanche-Olsen: On the structure and tensor products of JC-algebras, *Can. J. Math.*, to appear.

58 L. Harris; A generalization of C^*-algebras, *Proc. Lond. Math. Soc.* (3), **42** (1981), 331–61.

59 B. Iochum: Cones autopolaires et algebres de Jordan, These, Provence, 1982.

60 B. Iochum and F. W. Shultz: Normal state spaces of Jordan and von Neumann algebras, *J. Funct. Anal.*, **50** (1983), 317–28.
61 R. Iordanescu: Jordan algebras with applications, Preprint, 1979.
62 N. Jacobson: Macdonald's theorem on Jordan algebras, *Arch. Math.*, **13** (1962), 241–50.
63 N. Jacobson and C. E. Rickart: Jordan homomorphisms of rings, *Trans. Am. Math. Soc.*, **69** (1950), 479–502.
64 G. Janssen: Formal-reelle Jordanalgebren unendlicher Dimension und verallgemeinerte positivitätsbereiche, *J. Reine Angew. Math.*, **249** (1971), 173–200.
65 G. Janssen: Reelle Jordanalgebren mit endlicher Spuhr, *Man. Math.*, **13** (1974), 237–73.
66 G. Janssen: Die Struktur endlicher schwach abgeschlossener Jordanalgebren. Teil I. Stetige Jordanalgebren, *Man. math.*, **16** (1975), 277–305.
67 G. Janssen: Die Struktur endlicher schwach abgeschlossener Jordanalgebren. Teil II. Diskrete Jordanalgebren, *Man. Math.*, **16** (1975), 307–32.
68 G. Janssen: Factor representations of type I for noncommutative JB- and JB*-algebras, Preprint, 1982.
69 P. Jordan, J. von Neumann and E. Wigner: On an algebraic generalization of the quantum mechanical formalism, *Annals Math.*, **35** (1934), 29–64.
70 R. V. Kadison: Isometries of operator algebras, *Annals Math.*, **54** (1951), 325–38.
71 R. V. Kadison: A representation theory for commutative topological algebra, *Mem. Am. Math. Soc.*, **7** (1951).
72 R. V. Kadison: A generalized Schwarz inequality and algebraic invariants for operator algebras, *Annals Math.*, **56** (1952), 494–503.
73 W. Kaup: Algebraic characterization of symmetric complex Banach manifolds, *Math. Ann.*, **228** (1977), 39–64.
74 W. Kaup: Jordan algebras and holomorphy, Lectures at Univ. Fed. de Rio de Janeiro.
75 W. Kaup: Contractive projections on Jordan C^*-algebras and generalizations, *Math. Scand.*, to appear.
76 W. Kaup and H. Upmeier: Jordan algebras and symmetric Siegel domains in Banach spaces, *Math. Z.*, **157** (1977), 179–200.
77 M. Koecher: Positivitätsbereichen in $\mathbb{R}^m$, *Am. J. Math.*, **79** (1957), 595–6.
78 I. G. Macdonald: Jordan algebras with three generators, *Proc. Lond. Math. Soc.* (3), **10** (1960), 395–408.
79 K. McCrimmon: Jordan algebras and their applications, *Bull. Am. Math. Soc.*, **84** (1978), 612–27.
80 K. McCrimmon: Macdonald's theorem with inverses, *Pacific J. Math.*, **21** (1967), 315–25.
81 J. von Neumann: On an algebraic generalization of the quantum mechanical formalism, I, *Math. Sbornic*, **1** (1936), 415–84.
82 G. K. Pedersen: Monotone closures in operator algebras, *Am. J. Math.*, **94** (1972), 955–61.
83 G. K. Pedersen and E. Størmer: Traces on Jordan algebras, *Can. J. Math.*, **34** (1982), 370–3.

84 A. G. Robertson: Automorphisms of spin factors and the decomposition of positive maps, *Q. J. Math. Oxford* (2), **34** (1983), 87–96.

85 A. G. Robertson: Positive extensions of automorphisms of spin factors, *Proc. R. Soc. Edin.*, **94A** (1983), 71–77.

86 A. G. Robertson and M. A. Youngson: Positive projections with contractive complements on Jordan algebras, *J. Lond. Math. Soc.* (2), **25** (1982), 365–74.

87 D. W. Robinson and E. Størmer: Lie and Jordan structure in operator algebras, *J. Austr. Math. Soc.* A, **29** (1980), 129–42.

88 I. E. Segal: Postulates for general quantum mechanics, *Annals Math.*, **48** (1947), 930–48.

89 S. Sherman: On Segal's postulates for general quantum mechanics, *Annals Math.*, **64** (1956), 593–601.

90 S. Shirali: On the Jordan structure of complex Banach *algebras, *Pacific J. Math.*, **27** (1968), 397–404.

91 F. W. Shultz: On normed Jordan algebras which are Banach dual spaces, *J. Funct. Anal.*, **31** (1979), 360–76.

92 F. W. Shultz: Dual maps of Jordan homomorphisms and *-homomorphisms between C^*-algebras, *Pacific J. Math.*, **93** (1981), 435–41.

93 A. M. Sinclair: Jordan homomorphisms and derivations of semisimple Banach algebras, *Proc. Am. Math. Soc.*, **24** (1970), 209–14.

94 A. M. Sinclair: Jordan automorphisms on a semisimple Banach algebra, *Proc. Am. Math. Soc.*, **25** (1970), 526–35.

95 R. R. Smith: On non-unital Jordan–Banach algebras, *Math. Proc. Camb. Phil. Soc.*, **82** (1977), 375–80.

96 P. J. Stacey: Real structure in the approximately finite dimensional II_∞ factor, Preprint, Melbourne, 1981.

97 P. J. Stacey: Type I_2 JBW-algebras, *Q. J. Math. Oxford* (2), **33** (1982), 115–27.

98 P. J. Stacey: Locally orientable JBW-algebras of complex type, *Q. J. Math. Oxford* (2), **33** (1982), 247–51.

99 P. J. Stacey: The structure of type I JBW-algebras, *Math. Proc. Camb. Phil. Soc.*, **90** (1981), 477–82.

100 P. J. Stacey: Local and global splittings in the state space of a JB-algebra, *Math. Annalen*, to appear.

101 P. J. Stacey: Real structure in sigma-finite factors of type III_λ $(0<\lambda<1)$, *Proc. Lond. Math. Soc.* (3), **47** (1983), 275–84.

102 E. Størmer: On the Jordan structure of C^*-algebras, *Trans. Am. Math. Soc.*, **120** (1965), 438–47.

103 E. Størmer: Jordan algebras of type I, *Acta Math.*, **115** (1966), 165–84.

104 E. Størmer: On antiautomorphisms of von Neumann algebras, *Pacific J. Math.*, **21** (1967), 349–70.

105 E. Størmer: Irreducible Jordan algebras of self-adjoint operators, *Trans. Am. Math. Soc.*, **130** (1968), 153–66.

106 E. Størmer: On partially ordered vector spaces and their duals, with applications to simplexes and C^*-algebras, *Proc. Lond. Math. Soc.*, **18** (1968), 245–65.

107 E. Størmer: Real structure in the hyperfinite factor, *Duke Math. J.*, **47** (1980), 145–53.

108 E. Størmer: Decomposition of positive projections on C^*-algebras, *Math. Ann.*, **247** (1980), 21–41.

109 E. Størmer: Positive projections with contractive complements on C^*-algebras, *J. Lond. Math. Soc.* (2), **26** (1982), 132–42.

110 D. M. Topping: Jordan algebras of self-adjoint operators, *Mem. Am. Math. Soc.*, **53** (1965).

111 D. M. Topping: An isomorphism invariant for spin factors, *J. Math. Mech.*, **15** (1966), 1055–63.

112 H. Upmeier: Derivation algebras of JB-algebras, *Man. Math.*, **30** (1979), 199–214.

113 H. Upmeier: Derivations of Jordan C^*-algebras, *Math. Scand.*, **45** (1980), 251–64.

114 H. Upmeier: Automorphism groups of Jordan C^*-algebras, *Math. Z.*, **176** (1981), 21–34.

115 H. Upmeier: Derivations and automorphisms of Jordan C^*-algebras, *Proc. Symp. Pure Math.*, **38,** 2 (1982), 291–6.

116 J. D. M. Wright: Jordan C^*-algebras, *Mich. Math. J.*, **24** (1977), 291–302.

117 J. D. M. Wright and M. A. Youngson: A Russo Dye theorem for Jordan C^*-algebras, in K.-D. Bierstedt and F. Fuchssteiner (eds), *Functional Analysis: Surveys and Recent Results,* Proc. Conf. Paderborn, 1977, pp. 279–82.

118 J. D. M. Wright and M. A. Youngson: On isometries of Jordan algebras, *J. Lond. Math. Soc.* (2), **17** (1978), 339–44.

119 A. Wulfsohn: Tensor products of Jordan algebras, *Can. J. Math.*, **27** (1975), 60–74.

120 M. A. Youngson: A Vidav theorem for Banach Jordan algebras, *Math. Proc. Camb. Phil. Soc.*, **84** (1978), 263–72.

121 M. A. Youngson: Equivalent norms in Banach Jordan algebras, *Math. Proc. Camb. Phil. Soc.*, **86** (1979), 261–9.

122 M. A. Youngson: Hermitian operators on Banach Jordan algebras, *Proc. Edinb. Math. Soc. II*, **22** (1979), 169–80.

123 M. A. Youngson: Non-unital Banach Jordan algebras and C^*-triple systems, *Proc. Edinb. Math. Soc. II*, **24** (1981), 19–29.

124 M. A. Youngson: Completely contractive projections on C^*-algebras, *Q. J. Math.*, to appear.

125 H. Behncke: Finite dimensional representations of JB-algebras, *Proc. Amer. Math. Soc.*, **88** (1983), 426–28.

126 L. J. Bunce: The ordered vector space structure of JC-algebras, *Proc. Lond. Math. Soc.* (3), **22** (1971), 359–68.

127 L. J. Bunce: The theory and structure of dual JB-algebras, *Math. Zeit.*, **180** (1982), 514–25.

128 L. J. Bunce: Type I JB-algebras, *Quart. J. Math. Oxford* (2), **34** (1983), 7–19.

129 L. J. Bunce: A Glim–Sakai theorem for Jordan algebras, *Quart. J. Math. Oxford* (2), **34** (1983), 399–405.

130 B. Iochum: Non-associative L^p spaces, Preprint, 1983.

Index